CHEMISTRY RESEARCH AND APPLICATIONS

COMPUTATIONAL CHEMISTRY

THEORIES, METHODS AND APPLICATIONS

CHEMISTRY RESEARCH AND APPLICATIONS

Additional books in this series can be found on Nova's website under the Series tab.

Additional e-books in this series can be found on Nova's website under the e-book tab.

CHEMISTRY RESEARCH AND APPLICATIONS

COMPUTATIONAL CHEMISTRY

THEORIES, METHODS AND APPLICATIONS

DARIA BOVE
EDITOR

nova publishers
New York

For permission to use material from this book please contact us:
Telephone 631-231-7269; Fax 631-231-8175
Web Site: http://www.novapublishers.com

Library of Congress Cataloging-in-Publication Data

Computational chemistry : theories, methods and applications / [edited by] Daria Bove.
pages cm. -- (Chemistry research and applications)
Includes bibliographical references and index.
ISBN: 978-1-63321-354-8 (hardcover)
1. Chemistry--Data processing. 2. Cheminformatics. I. Bove, Daria, editor.
QD39.3.E46C628 2014
541.0285--dc23

2014022559

Published by Nova Science Publishers, Inc. † New York

CONTENTS

PREFACE

This book begins with a discussion on the protein kinase targeting drug discovery and design. It continues with topics on computational chemistry investigation of UV filters; using the network simulation method to study ionic transport processes in electrochemical cells; molecular simulation of electron beam nanofabrication; interpreting of chiral ordering of hybrid system of several AZO dyes; a general procedure for a priori calculation of thermochemical properties of organic molecules and free radicals; strategies for design of new organic molecular rectifiers; and DFT studies.

Chapter 1 – The development of new drugs always was time consuming and costly. With the development in experimental methods was possible to scan small-compounds libraries in order to find potentially suitable molecules. Nevertheless, these methods are able to so with a very low rate. Protein kinases are a class of enzymes involved in the great majority of cellular process. Due to its presence in many signal pathways, cell cycle and gene expression control mechanisms, this class is one of the major targets for pharmaceutical industries today. Aiming the reduction of time and costs in drug development computational tools started to be used, and commonly called Computer aided drug design (CADD). These methods can be separated in two different types of approaches, structure based (SBDD) and ligand based (LBDD). In order to use structure based methods, information about the 3D structure of the target is needed, normally obtained through x-ray crystallography or NMR. Ligand based methods are preferred when this information is not available. The focus of the present chapter is the structure based methods used in the computer drug discovery and design process targeting kinases. Starting with the visual and energetic analysis of the binding site, it is obtained enough information for the creation of a Pharmacophore model and the application of Virtual Screening process. The results of the screening can then be analysed through Molecular Docking process followed by Molecular Dynamics in order to better simulate the real binding conditions. The obtained complexes can be analysed using energy calculations, in order to better understand the binding process. Within the described procedures, computational tools originate a great amount of useful information since hit discovery from lead optimization. The data obtained can save time and reduce costs in the process of drug design and discovery in a large scale.

Chapter 2 – The raising level of awareness regarding the harmful effects of solar radiation has resulted in an increase in the production and use of sunscreens. These commercial products contain ultraviolet (UV) filters that absorb, reflect or scatter UV radiation (290-320 nm for UVB and 320-400 nm for UVA) therefore preventing sunburn, photo-aging and ultimately skin diseases such as skin cancer. A high screening efficiency can only be guaranteed if a UV filter has a high photo-stability. However, UV filters are known to undergo degradation upon exposure to sunlight and chemical substances used in water disinfection such as chlorine, leading to the formation of undesirable byproducts with properties different from the parent compounds. This chapter presents a comprehensive review on the computational studies performed so far on the structural, electronic, energetic and UV absorption properties of some of the most widely used UV-filters and their degradation products (UVF-DP). *Ab initio* and density functional theory (DFT) methods have been used to obtain the molecular structure, energetics and other molecular properties of UVF-DP. UV-Vis absorption spectra and excited state properties have been calculated using the time dependent density functional theory (TD-DFT). The influence of different solvents in these properties has been studied using solution models. Transformation and degradation mechanisms have been established with the aid of computational methods and results indicate that these can be employed to assess the environmental fate of UV-filters.

Chapter 3 – The network simulation method is used to numerically study the steady-state, transient and small-signal ac responses of a binary electro-lyte solution placed between two electrodes. The ionic transport processes in the solution are described by the Nernst-Planck and Poisson equations, while the interfacial charge transfer processes are characterized by the Chang-Jaffé equations. The chronoamperometric, chronopotentiometric, linear sweep voltammetric, and small-signal ac responses of symmetric cells, the cation being the only electroactive species at both electrodes, are simulated on the basis of an only network model. The main advanta-ges of the network simulation method with respect to other electrochemi-cal numerical methods, are emphasized. They can be summarized as follows: (i) The network simulation method permits to impose any condition on the electric potential and the electric current; (ii) Steady-state, transient and small-signal ac responses can be dealt with by means of this approach on the basis of an only network model; and (iii) It permits the study of any transport process through spatial regions where large gradients in the ionic concentrations and the electric potential occur.

Chapter 4 – Electron beam irradiation is expected to become a technique for tailoring the structure of materials to obtain desirable properties. Molecular dynamics (MD) simulations are a powerful tool to understand the structural changes of materials at the atomic scale. In this chapter, MD simulations of electron beam nanofabrication are introduced. Electron beam modification of carbon nanomaterials, the pattern formation process in electron beam lithography, and electron irradiation damage in silicon oxides are investigated with MD simulations.

Chapter 5 – It is well known that Weigert effect emerges optical anisotropy induced by linearly polarized light in the case of azo dyes, liquid crystals, and such polymers and its application for holography. We have reported on supramolecular interactions for controlling molecular arrangement of photo-inactive metal complexes between directly photo-functional azo dyes and photo-inactive metal complexes in flexible polymer films. In this work, the authors have prepared PMMA cast films of three azo dyes, such as azobenzene (**AZ**), disodium 6-hydroxy-5-((2-methoxy-5-methyl-4-sulfophenyl)azo)-2-naphthalenesulfonate (Allura Red), 4,4'-di-n-octyloxyazoxybenzene, and hybrid materials containing a new chiral Schiff base Co(II) complex ($C_{32}H_{32}CoN_2O_2$ showing IR band (C=N) at *c.a.* 1630 cm^{-1}) as a chiral dopant. After circularly polarized UV light irradiation, they compared and found that the degree of chiral ordering (increasing of CD intensity) of **AZ** was the highest among three ones. 4,4'-Di-*n*-octyloxyazoxybenzene having long chain exhibited characteristic increasing of dichroism at $\pi-\pi^*$ band, which may affect on chiral ordering. With the aid of computational chemistry, the authors attempted to interpret interactions between transition dipole moments of azo dyes (especially **AZ** was effective) and Co(II) complexes as their light-induced helical arrangement. Theoretical treatment with ZINDO calculations of spectra about the angles of transition dipole moments suggested that supramolecular interactions induced by polarized light predominantly resulted from $\pi-\pi^*$ transition for **AZ**+PMMA, whereas n-π^* transition for Co(II) complex+**AZ**+PMMA.

Chapter 6 – There have been many attempts in the literature to use variations of the "3rd law" method for estimating several thermodynamic properties (including the entropy and the heat capacity) for a variety of molecules. Application of the method to cases of free radicals is much less common. This is because of the difficulties and uncertainties associated with the assignment and the calculation of the parameters required for providing the higher order anharmonicity related corrections to improve upon the values of the properties obtained under the basic rigid-rotor harmonic-oscillator (RRHO) approximation.

The study reported in this chapter devised and standardised a procedure based on the consistent theoretical premises to make a priori estimates of ideal gas entropy and heat capacity of organic molecules and especially free radicals with acceptable accuracies. An important part of the procedure is to calculate in a transparent and scalable manner the higher order corrections due to hindered rotation (HR) of groups internal to the molecular species to be added to their RRHO approximated property values. Their HR correction procedure adapts the classical Pitzer method to a range of large and branched radicals containing multiple rotating tops, either symmetric or asymmetric, with single or repeat occurrences of a given top within the same radical. Although a commercially available software (such as Gaussian) may allow a user an automatic way to calculate the corrections of the property values due to hindered rotations (HR) at some low vibrational frequencies, the authors experience suggests that caution should be exercised before accepting the software output in this regard. The corrected property values for a number of different classes of radicals, published by previous researchers, were found to be more closely reproduced by using the methodology presented in this chapter than using Gaussian software's HR correction facility. For larger molecules and radicals that may exist as multiple low-energy stable conformers, an additional correction due to the presence of a mixture of conformers was used in an approximate manner for entropy only. This correction is expected to make the prediction of entropy more accurate. Gaussian calculation does not seem to account for the conformers.

Chapter 7 – Despite many efforts to date, the application of molecular rectification has been, and remains, very limited vis-à-vis use of traditional silicon-based rectifiers. In order to render the latter a more realistic proposition and to deliver tangible progress in this respect, a greater understanding of fundamental underlying mechanisms from a chemical perspective becomes sine qua non; indeed, molecular simulation has much to offer such an understanding. The three most common kinds of molecular rectifiers composed of Donor-σ-Acceptor, Donor-π-Acceptor or Donor-Acceptor molecular systems, and their chemistry plays a vital role through the synthetic versatility of the functionality of molecular rectifiers. The authors have reviewed substitution, conformation, molecular length, anchoring, annulation, and isomeric effects as most common structural motifs used to design molecular rectifiers. The nature of the electrode and the external applied electric field also has a direct impact on rectification. The advent of experimental progress in molecular rectifiers has given rise new challenges to theory in developing theoretical tools to describe the electron transport in molecular rectifiers. Up to now, the combination of Density Functional Theory (DFT) and non-equilibrium Green's function (NEGF) methods have been widely used to study the quantum transport through nanoscale devices. Here, they offer a perspective on how molecular simulation approaches can be used to improve and inform design strategies for molecular rectifiers.

Chapter 8 – The 8π,6π-electrocyclization processes of substituted conjugated tetraenes and pentaenes were investigated by quantum-chemical calculations - density functional (DFT) methods. Reactivity of substituted octatraenes with double bond incorporated into the benzene ring in which substituents are positioned at various position(s) of conjugated polyenes was studied. Influence of terminal substituents on reaction mechanism and the product *endo/exo* selectivity was also established.

In: Computational Chemistry
Editor: Daria Bove

ISBN: 978-1-63321-354-8

Chapter 1

PROTEIN KINASE-TARGETING DRUG DISCOVERY AND DESIGN: COMPUTATIONAL CHEMISTRY AS AN INDISPENSABLE TOOL

Pedro M. M. Araújo, Luís Pinto da Silva and Joaquim C. G. Esteves da Silva*

Centro de Investigação em Química, Departamento de Química e Bioquímica, Faculdade de Ciências da Universidade do Porto, Porto, Portugal

ABSTRACT

The development of new drugs always was time consuming and costly. With the development in experimental methods was possible to scan small-compounds libraries in order to find potentially suitable molecules. Nevertheless, these methods are able to so with a very low rate. Protein kinases are a class of enzymes involved in the great majority of cellular process. Due to its presence in many signal pathways, cell cycle and gene expression control mechanisms, this class is one of the major targets for pharmaceutical industries today. Aiming the reduction of time and costs in drug development computational tools started to be used, and commonly called Computer aided drug design (CADD). These methods can be separated in two different types of approaches, structure

* Corresponding author: Joaquim C.G. Esteves da Silva, e-mail: jcsilva@fc.up.pt; Tel: (+351) 220-402-569; Fax: (+351) 220-402-659.

based (SBDD) and ligand based (LBDD). In order to use structure based methods, information about the 3D structure of the target is needed, normally obtained through x-ray crystallography or NMR. Ligand based methods are preferred when this information is not available. The focus of the present chapter is the structure based methods used in the computer drug discovery and design process targeting kinases. Starting with the visual and energetic analysis of the binding site, it is obtained enough information for the creation of a Pharmacophore model and the application of Virtual Screening process. The results of the screening can then be analysed through Molecular Docking process followed by Molecular Dynamics in order to better simulate the real binding conditions. The obtained complexes can be analysed using energy calculations, in order to better understand the binding process. Within the described procedures, computational tools originate a great amount of useful information since hit discovery from lead optimization. The data obtained can save time and reduce costs in the process of drug design and discovery in a large scale.

1. INTRODUCTION

The procedure to develop and uncover new drugs was always time consuming. Before the automation of the processes, the manual synthesis followed by the test of small groups of compounds was extremely inefficient. With the advances in chemistry linked to the emergence of new techniques as combinatorial chemistry and high-throughput screening (HTS) was finally possible to synthesized and analyse a large number of compounds within relatively small periods of time. With the progress in the number of chemical compounds screened was expected an increase in the number of new drugs obtained. However, even with the growth in new drug research funding, the results were not proportionally. From 1993 to 2007 the investment in this area almost quadrupled. However, in the same period of time, the number of new approved drugs did not suffer a significant improvement. That is mainly a consequence of the high failure rate verified, as in 10 000 tested molecules only 1 or 2 have the potential to reach the market. The cost from a drug discovery and development program from the start to the implementation of the product in the market can take up to 14 years with a cost near 1 billion USD (United states dollar) [1–3].

In the post-genome era the amount of available information increased exponentially. Combining that with the high cost of experimental procedures created the need for a more efficient approach to the topic. Since it is impossible to treat efficiently huge quantities of data manually, automated analysis procedures naturally emerged as an answer to the challenge (Figure 1). These methods are known by Computational tools or more commonly as Bioinformatics tools [4].

Virtual high-throughput screening (vHTS) or only Virtual screening (VS), can be classified as the class of computer tools able to automatically evaluate very large libraries of compounds [5]. Doman et al. performed a project to identify novel inhibitors for the enzyme protein tyrosine phosphatase 1B (PTP1B), pointed as a key factor in type II diabetes [6]. In this study both HTS and VS were used. The experimental HTS was performed in a library of 4,000,000 compounds while VS was applied to 235,000 molecules using the X-ray structure of PTP1B. The hit rate of the experimentally method was 0.021% with only 85 compounds showing IC50 values inferior to 100 μM. On the other side, the computational method obtained 365 high-scoring molecules of which 127 (34.8%) had IC50 value inferior to 100 μM in the *in vitro* tests. The best IC50 result was obtained using the vHTS methodology with the value of 1.7 μM, when the best result for classic HTS was 4.2μM. This work gives us a rare comparison between HTS and VS and elucidates the potential of the computational tools.

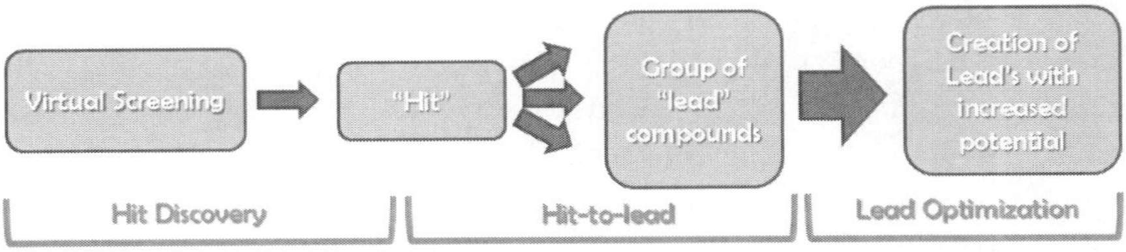

Figure 1. Schematic representation of the Hit discovery to lead optimization process.

Computational methods rapidly became core tools in the novel drug discovery and development, being commonly called Computer aided drug design (CADD) [7]. It is impossible to deny the contribution of these methods in drug development programs. Computational methods can simulate almost all the aspects in the drug discovery and design process [3]. However, they cannot stand alone in the creation of novel drugs [8]. CADD is especially useful in reducing the number of compounds needed to be experimentally tested. The activity prediction allows the exclusion of the some molecules while focusing on those forecasted to be active, resulting in a more time efficient "hit identification". A hit is a compound with specificity and potency to bind to the desired target [9]. The next step is the so called "hit-to-lead" phase, where a lead series is created based on the interesting characteristics of the most promising hit [10]. The selected molecules undergo an improvement process called "lead optimization". Due to the high costs of this phase the usage of computational tools in it show a big economic impact. CADD is able to point the way that lead optimization should take, and also to save time and money that will be needed to explore several options. Computational aided drug design enclosures two big subgroups of methods. Structure-based (SBDD) and Ligand-Based (LBDD). SBDD is highly dependent on the 3D information of the target, normally obtained through x-ray crystallography or NMR. Structure based methods aim the creation of specific molecules to the binding site of the target. LBDD focus on the information available on the active molecules known to bind to the target, searching for compound with chemical similarities. Ligand based methods are preferred when the available information on the structure of the target is not enough to perform SBDD [2, 11].

Kinases have the ability to transfer a phosphate groups manly from ATP or GTP to cellular targets as proteins, nucleotides and lipids. Protein kinases are able to phosphorylate proteins at Ser, Thr or Tyr residues, being the most significant subgroup of kinases. Protein phosphorylation is abundant within the Eukaryotic cell, being estimated that at a given time one third of all the cellular proteins are phosphorylated. As a consequence of its abundance, kinases can be found in almost all cellular mechanisms but especially in the regulation of signal transduction pathways [12–14]. In recent years kinases have become the central target in cancer drug design. This attention is manly a result of the success of Imatinib, also known as Gleevs. With the ability to inhibit a variety of tyrosine kinases this small molecule has been used in the treatment of diverse types of cancer. However, Imatinib had more relevance in the treatment of chronic myelogenous leukemia (CML) and gastrointestinal tumours (GIST). The effect of this compound was tremendous in transforming these two extremely fatal cancers in "possible to control" conditions. Nevertheless, protein kinase inhibitors have more therapeutic targets that cancer treatments, having already been approved kinase inhibitors in the treatment of inflammatory diseases [15].

During the present chapter we aim to show the common approach to a rational drug design pipeline using structure-based methods while providing some examples of application in the study of protein kinases.

2. Binding Site Identification

Ideally a CADD program begins with the existing 3D structure of the target protein bound to a ligand, obtained through x-ray crystallography or NMR. However, the 3D structure of the target alone also represents a good starting point. When the data is not available it is possible to use Homology modelling processes, based on the belief that similar amino acid sequences originate similar three dimensional protein structures [16]. When a bonded ligand is absent and the binding site is unknown the first challenge is to define the binding site. The ligand binding site is generally a depression in the protein surface. The binding process occurs when in a defined site there are specific interactions between the protein and the ligand, generally non-covalent, and those are superior in strength to the repulsive contributions. The non-covalent interactions can be placed in four categories: hydrogen bonds, Van der Waals forces, π-effects and ionic interactions [17]. Computational approaches can be used to identify and characterize high-affinity or new binding sites. These methods are usually divided in several distinct classes: geometric methods, energetic based methods, combination of the previous two, pocket matching and molecular dynamics (MD) based detection.

Geometric methods use grids to identify cavities or to define the 3D structure of a protein. Examples of programs using this strategy are POCKET [18], LIGSIET [19] and its extension LIGSITEcsc [20], SURFNET [21], ConCavity [22], APROPOS [23] and DEPTH [24]. Energetic-based methods are more sensitive and specific that geometric methods using the same amount of time. These methods use a probe system to identify favourable interactions in the protein surface, and collections of positive points represent possible binding sites. Programs using energetic calculations to evaluate binding sites are Q-Site Finder [25], SITEHOUND [26], POCKETPICKER [27], one developed by Morita el al [28] and FLASPSITE [29]. The downside of these two classes of methods result from the large number of false-positive results originated [30]. Pocket matching uses data from known binding sites to evaluate the existence of these regions in other proteins. This method follows the principle that binding sites have unique characteristics when compared to the other protein regions. There are some methods based on this idea like Catalytic Site Atlas [31], AFT [32], Pocket-surfer [33] and Patch-surfer [34]. Proteins are not fixed molecules. They present some fluctuation between conformations. Sometimes the available molecular conformation is not the ideal to the binding of other molecules due to the shape of the binding site. In these cases a molecular dynamic step needs to be added to the process in order to obtain a set of multiple conformations of the target. It is more likely to find the ideal conformation of a binding site in a group or conformations that in only one static structure. For more information on molecular dynamics see the respective topic (further on this chapter).

The previously presented methods can identify correctly the binding site in 70-90% of the cases when the protein is in the ideal binding conformation (holo) but only in 50-70% if the protein is not in this conformation (apo) [11, 35]. The obtained binding sites results can them be used to guide a Virtual screening process (VS) in order to find new possible ligands.

3. Virtual Screening

Virtual screening (VS) is the computational analogue of the high throughput screening (HTS), and therefore can also be named virtual high throughput screening (VHTS). To avoid misunderstandings on this chapter we refer to this technique only as Virtual screening (VS). It can be used to evaluate libraries that contain a vast number of compounds in an automatically and consequently fast mode. The purpose of this method is not to obtain a high number of molecules but otherwise achieve novel structures with the desired pharmacological profiles, called virtual screening hits [5, 36].

VS is undoubtedly a core technique in rational drug discovery and design programs, both for academic research and pharmaceutical companies [37]. Its value raised with the constant growth in the number and size of compounds databases, especially virtual databases [5, 36]. Virtual screening can evaluate a large number of compounds more efficiently in a less time consuming and less expensive manner then HTS, being preferred for the study of bigger databases. *In silico* methods consume computational time in order to decrease the costs of the experimental phase of the study. Nevertheless, VS does not exclude HTS. In fact the use of both methods together resulted in the prompt identification of novel compounds [36]. VS can be seen as a method to restrict the number of residues in the study before HTS, decreasing expenses and saving time.

In Structure Based VS (SBVS), also called Receptor Based VS, the search aims specific features observed in the receptor-ligand interaction [38], such as hydrogen bound acceptors or donors, polar or apolar regions and particular conformations. Its objective is to differentiate ligands that bind strongly to the target from those who not. However, the search can be focused in characteristics independent of the target as drug-likeness [36]. Methods based in ligands characteristics are commonly called Ligand based Virtual Screening

(LBVS). Thus, VS can be divided in Structure based (SBVS) and Ligand based (LBVS) methods.

Structure based methods can be further divided in two different classes, active site derived pharmacophore methods and molecular docking.

3.1. Pharmacophore

The concept of Pharmacophore was describe by IUPAC [39] as "an ensemble of steric and electronic features that is necessary to ensure the optimal supramolecular interactions with a specific biological target and to trigger (or block) its biological response" [40].

The creation of a pharmacophoric model can be achieved with the inclusion of the relevant binding characteristics observed in the 3D structure of the receptor, specially focusing the points that are already known to create interactions to molecular targets. Hydrogen bond acceptors or donors and charged groups are the main features to look for. Pharmacophore methods focus exclusively in the binding site and its features, a simplification that decreases the computational demand in its usage.

Pharmacophore models can be applied in different stages of the novel drug discovery and design process, as de novo design and lead optimization but excelling in virtual screening [11, 38]. The usage of pharmocophore models for virtual screening of compounds databases sometimes face the limitation in the number of features that can be used in the process, with the maximum of seven, although more than these were found in the target analysis. More screening parameters mean more computational time spent. In order to study one target it is ideal to study more than one 3D structure of it. However this is not always possible due to the lack of 3D representations [40, 41].

It is possible to create pharmocophore models from the 3D structure of the target alone. However, the usage of the complex target-ligand is the most common and the ideal starting point. Naturally structure based pharmocophore methods are divided in complex (with ligand) based and macromolecular (without ligand) based.

In the complex based approach the binding site is already known and the challenge is to find the central interactions between the target and the existing ligands. Without available 3D structures of the target-ligand complex this approach is unavailable. Ligand Scout [42] and Pocket v.2 [43] are two examples of programs that perform this type of analysis.

In the absence of the 3D structure of the complex but with existing structure of the target, macromolecular based methods can be applied. If the binding site is known its study can begin, if not it needs to be defined (see section Binding site identification). HS-Pharm [44] and the protocol developed by Tintori et al. [45] are examples of programs that use macromolecular structure based pharmocophore approaches.

3.2. Molecular Docking

Docking tries to preview the position and orientation of two different molecules at the time they interact. When a small molecule is docked in a target protein the process is commonly named protein-ligand docking [46, 47]. These methodologies rely on the docking protocol (Figure 2) and in the quality of the used structures to obtain high value hits, a model independent of any human subjective evaluation [48]. Protein-ligand docking is the "number one" choice in virtual screening protocols, as it generally creates extra and more detailed information that than the other structure based or ligand based methods. It is also computationally more expensive [38].

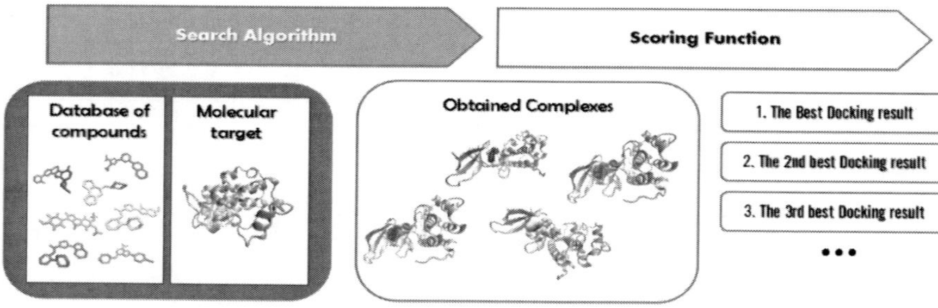

Figure 2. Schematic representation of the Molecular Docking protocol.

There are currently many docking software packages, and due to the relevance of these types of programs they are in constant update with new versions being released within short periods of time. AutoDock4 [49] is the most recent version of this family of programs and allows ligand flexibly and at the same time limited receptor flexibility. Its release was made together with the suit of programs AutoDockTools4, a user friendly interface for the usage of AutoDock. AutoDock Vina is another docking program package similar to AutoDock but with improvements in the speed and accuracy of the calculation thanks to the parallelism using multithreading on multicore machines [50]. Genetic Optimisation for Ligand Docking (GOLD), uses a genetic algorithm that allows the complete flexibly of the ligand and partially of the receptor while being able to displace water molecules on the binding site [51]. Any docking protocol has two main components: the search algorithm and the scoring function.

3.2.1. Search Algorithm

The search algorithm, sometimes also called docking algorithm, explores the possible positions and orientations of the ligand within the target. In order to perform the docking in a fast pace and being able to rapidly dock large libraries of compounds, the searching algorithms cannot be too exigent in computational power. However, the needed degrees of freedom to find the true binding mode must be included. Thus, the docking algorithm always tries to create the best ratio between the effectiveness in screening the binding site and the speed of the process [38]. Flexibility of both, the ligand and the protein, needs to be taken into account. The available 3D structures of the target do not necessarily represent the binding state of the target or the conformation of its binding site when occupied, as proteins are dynamic molecules in constant motion specially when interacting with other structures. Although in some rigid proteins there is a dominant conformation, in the majority of the cases it does not happen [46, 52].

Flexibility is accepted as one of the major challenges in docking. To simulate structures flexibility, more degrees of freedom must be included in the calculation making it more time consuming. In the beginning of protein-ligand docking, flexibility was not taken into account, and both the ligand and the protein were treated as rigid structures. This is the most basic approach and consequently the faster. Consequently the rate of false negatives created was high with the loss of many hit compounds that failed to bind to the single conformation of the available binding site [38]. The ideal protocol would always consider ligand and target flexibility, however due to the computational demand of it, today the most common approach considers only the flexibility of the ligand. In order to bypass this issue some approaches have been delineated like the "hierarchical docking" in which the faster method is used previously to decrease the amount of compounds in the study, and afterwards the more accurate method could be applied for a smaller pool of compounds [38, 53]. Still, the usage of protocols like this need to give special attention to the creation of false negative results from methods used early like rigid protein-ligand docking that would exclude potential interesting residues prematurely.

3.2.2. Scoring Functions

After the screening of the target binding site the obtained conformations need to be evaluated by the scoring function. Is not enough to find the proper posing, it is crucial to successfully identify it. Scoring functions need to be able to recognize the active compounds and distinguish the true binding posing from the others. The usage of an extremely rigorous scoring function would have increased computational time, turning the virtual screening of larger compound databases impracticable. In order to decrease the cost of the scoring function, part of its accuracy is sacrificed applying simplifications to the functions.

The ratio between accuracy and speed for the scoring functions is another of the major problems of docking [38]. The success rate of both the scoring functions and the search algorithm are highly dependent of the protein and the ligand in the study. It is expected that the same pair of scoring function and search algorithm achieve different accuracy rates by studying different targets, since both these functions are highly dependent on the molecular characteristics of the targets. There is not such a thing as a universal best pair of search function and scoring algorithm but there are in fact best pairs for different situation.

Generally the scoring functions can be classified according to the used method into three different classes: the force-field-based, the empirical-based and the knowledge-based. In the first class, force-field-based is used to assess the protein-ligand binding. These methods are computationally challenging since they do not enclose any experimental parameters to simplify the calculation. Nevertheless, it has known limitations like the inexact treatment of long range effects.

Empirical methodologies are scoring functions design to mimic experimental binding energy data. As a consequence these methods are highly dependent in the availability of experimental data and its accuracy. The last class, in contrast to the previously exposed scoring functions, focus on the structures instead of the binding energies. This is the less computational demanding class. Due to its dependence in existing data even knowledge-based methods can face limitations [38].

4. MOLECULAR DYNAMICS

As previously said biomolecules, especially proteins, are dynamic entities. Each molecule can roam between a group of different conformations, of which some are more prevalent than others. However, the most common position for a protein alone can be different from the bound state and even differ for complexes with different ligands. In order to better understand the binding of molecules to protein targets Molecular dynamics (MD) simulations are usually applied, creating an ensemble of conformations available to study [11].

MD uses Newtonian physics to simulate atomic motions of desired molecular targets. Without this simplification the calculations of the quantum-mechanical motions of molecules like protein would be unreal to perform due to the exigency in computational time. The first step in this process is to estimate the forces acting in each atom of the system, which is normally accomplish using Equation 1 [54]. The forces in this equation are generally translated into a force-field that describes the molecular motions resulting from the forces applied in each atom of the system. When the forces applied in each atom of the system are calculated, Newton's laws of motion are applied changing the positioning of the atoms. Then the simulation time progress in a time step defined by the user and the process repeats itself until the desired simulation time is achieved [11, 55]. Molecular dynamics still faces some limitation mainly due to the high amount of computation power required to large dynamics and due to limitations in some of the force fields available. The progress in other fields of study is of major relevance to the improvement of MD [55].

GPUs (graphics-processing-units) are an example of recent developed technology that can be used to speed up Molecular dynamics calculations [55–57]. DE Shaw group created processor units specific for MD calculations, thus obtaining a super computer able to perform microseconds of simulations per day and making it possible to observe events as protein folding and unfolding and drug-binding [55, 58, 59].

The Molecular dynamics simulation program packages more popular are the AMBER [60, 61], CHARMM [62] and NAMD [63, 64]. These platforms are sometimes mistaken by the force-fields they apply since they usually have the same name [55].

$$E_{total} = \sum_{bonds} K_r\,(r - r_{eq})^2 + \sum_{angles} K_\theta(\theta - \theta_{eq})^2 + \sum_{dihedrals} \frac{V_n}{2}[1 + \cos(n\phi - \gamma)] + \sum_{i<j}\left[\frac{A_{ij}}{R_{ij}{}^{12}} - \frac{B_{ij}}{R_{ij}{}^{6}} + \frac{q_i q_j}{\epsilon R_{ij}}\right]$$

Equation 1. Atomic forces that generate molecular motion. Can be divided in covalent, bonds, angles and dihedrals angles energy's and non-covalent caused by Vander wall interactions and electronic charges.

5. Examples of Success

In October of 2007 researchers at Merck Pharmaceutical Company successfully obtained Isentress™ (raltegravir) the first clinical-approved HIV Integrase inhibitor, which was only possible due to the previously computational work of McCammon et al. [65]. This case is considered by many the first explicit example of the structure based methods application in the discovery of a new drug.

Thenceforth many other drug discovery programs benefited from computational tools, being this field in constant growth. The Kinases are one of the largest families of human proteins, representing 1.7% of the total of this class [66]. The phosphorylation process can be seen as a molecular "on-off" mechanism, able to activate or inactive cellular effectors. The transfer of a phosphate group to an aminoacid of a protein can be enough to trigger a signal pathway resulting in a huge variety of cellular events since cell cycle abrogation/progression, from transcription of specific DNA regions to cellular apoptosis or survival mechanisms [66]. Alterations in protein kinases regulation has been related to many pathological events as diabetes, inappropriate inflammatory response, oncological diseases and others [66, 67]. Protein kinases gained special attention from the pharmaceutical industry in the search for cancer treatment drugs after the success of Imatinib (Glivec), a "blockbuster" drug that transformed Chronic Myelogenous Leukemia (CML) and Gastrointestinal Tumors (GIST) from fatal conditions with very bad prognostics into treatable situations. Is interesting to notice that in this particular case the low specificity of kinase inhibitors (affecting many protein kinases) was a positive factor that allowed these drugs to be so effective and at the same time treat more than one pathology. At the end of 2013 there was only 26 approved protein kinase inhibitors for clinical use, being 23 of them related to cancer treatment [15].

Table 1. Approved protein kinase inhibitors for clinical use [15]

Name	Target	Year of approval
Imatinib	CML	2001
Getitinib	Lung cancer	2005
Erlotinib	Lung, pancreatic and others cancers	2005
Sorafenib	Renal cancer	2005
Desatinib	CML	2005
Sunitinib	Renal cancer and GIST	2006
Temsirolimus	Renal cell carcinoma	2007
Nilotinib	CML	2007
Everolimus	Several cancers	2009
Lapatinib	Renal cancer	2009
Toceranib	Canine mastocytma	2009
Pazopanib	Renal cancer	2009
Cabozantinib	Canine thyroid cancer	2010
Masivet Kinavet	Canine mastocytoma	2010
Ruxolitinib	Myelofibrosis	2011
Crizotinib	NSCLC	2011
Vemurafenib	Melanoma	2011
Vadetinib	Thyroid cancer	2012
Axitinib	Renal cell carcinoma	2012
Bosutinib	CML	2012
Tivozanib	Kidney cancer	2012
Regorafenib	Thyroid cancer	2012
Lenvatinib	Thyroid cancer	2012

Depending on the site and mechanism of interaction protein kinases inhibitors can be classified in three different types. The first type contains the compounds that bind to the ATP binding site of the target, a conserved region along protein kinases. This is the standard class with a variable specificity that can go from low to high. Type II interact with the “extended ATP binding site”, a region less conserved than the ATP binding site making this class more target specific. Type III inhibitors don’t compete with ATP having particular binding sites in order regions of the target kinase, being the class more distant from the others [66].

Due to the high number of kinases that the human genome codes is sometimes difficult to define which effector is responsible to the phosphorylation of a particular target. To overcome this matter several computational tools have been created, and in 2014 at least two new features have been created, PKIS [68] and PSEA [69]. PKIS is a free web service reported as capable to outperform already exiting tools in the identification of protein kinases associated with phosphorylated targets as KinasePhos 2.0 [70], Musite [71], and GPS2.1 [72]. The comparison study between the different methods was made using data from Phospho.ELM [73], a database of serine, threonine and tyrosine phosphorylation sites. The authors of PSEA focused their attention in disease related phosphorylation subtracts, trying to preview which kinases are more likely to create abnormal phosphorylation events and which kinases can be related to known pathology. Remarkably, using this new method they were able to observe a correlation between MAPK (Mitogen-activated protein kinase) and GSK (Glycogen synthase kinase) families overexpression in diseases related to phosphorylation [69].

Casein kinase 2 (CK2) has a large pool of molecular targets, over 400, with distinct functions. Due to the high number of targets affected by this enzyme its overexpression can be related with pathologies as oncological diseases, deregulation of inflammatory response and others. The creation of inhibitors to CK2 was a must and it was only possible using virtual screening in the early stages of the process, followed by biochemical trials and chemical optimization. This process resulted in the creation of the first patent of inhibitors to CK2 by IMBG medicinal chemistry [66, 74, 75].

Janus kinase 2 and 3 (JAK2/3) are involved in lymphoid derivate diseases and have recently receiving attention from virtual screening structure based pharmocophore studies. The work of Rajeswari et al. [76] and Jsuja et al. [77] show some similarities in the used methods. Both authors created pharmocophoric profiles, followed by virtual screening and evaluation by molecular docking and other techniques. The purpose of both studies was also alike: create information to better understand the binding site of these two enzymes and identify novel molecules with inhibitory potential to these targets.

In a different approach Srinivasan et al. [78] studied Nek6, a NIMA (never in mitosis, gene A) related kinase and an important factor to the start of the cell cycle. This group performed a virtual screening study starting from a homology structure of the enzyme obtained by them. After the docking validation they reported the identification of two novel Nek6 inhibitors.

Focusing on the cell cycle study, especially related to the G2 checkpoint, our group studied Chk1 (Checkpoint kinase 1) [79]. This enzyme has a key role in the cell cycle control. To better understand the binding of potential inhibitors to this enzyme we started from already published structures of two compounds bond to Chk1, compound 39 (C39) and compound 40 (C40) [80], and we performed semi-empirical calculations [81] to obtain the free energy differences from the unbound state (Chk1 without ligand) to the complex state (Chk1 with ligand) and infer the energy contributions of 18 residues of the target binding site. We successfully identified the key factors in the binding of small molecules to Chk1 and simultaneously we showed that C39 had the best inhibitory potential in the study. This type of results can be further used in improvement programs targeting this protein kinase.

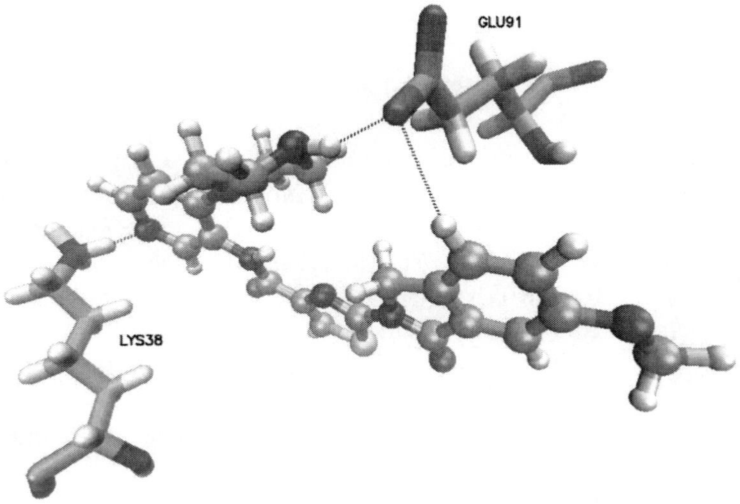

Figure 3. Representation of the interaction between Compound 39 and the two most favourable residues to its binding to Checkpoint kinase 1.

CONCLUSION

The usage of computational tools in the discovery and improvement of novel drugs is undoubtedly necessary. New drug discovery pipelines require high amounts of monetary investment and time. As such any method that is able to maintain the accuracy rate of the process, and at the same time make it faster and less costly must be embraced.

Structure-based methods are already well developed, and in a post-genomic era the number of available 3D structures of molecular targets will only grow, increasing the need for more efficient methods. Computational tools are always dependent on the developments related to hardware.

Nevertheless, more potent equipments will always require more advanced programs. This field of study will continue its growth and hopefully the challenges that we face today due to their computational exigency will be surpassed tomorrow and new objectives will rise.

The better understanding of protein kinases binding sites, its inhibitors and their functions is necessary to treat a high number of pathologies. Creating or finding molecules that bind to kinases in a specific way is still a challenge today but as it was reported in this chapter the efforts of researchers worldwide are creating small advances daily to a higher end.

REFERENCES

[1] Lobanov V. Using artificial neural networks to drive virtual screening of combinatorial libraries. *Drug Discov Today BIOSILICO* 2004;2:149–56.

[2] Ou-Yang S, Lu J, Kong X, Liang Z, Luo C, Jiang H. Computational drug discovery. *Acta Pharmacol Sin* 2012;33:1131–40.

[3] Shekhar C. In silico pharmacology: computer-aided methods could transform drug development. *Chem Biol* 2008;15:413–4.

[4] Yu U, Lee SH, Kim YJ, Kim S. Bioinformatics in the post-genome era. *J Biochem Mol Biol* 2004;37:75–82.

[5] Walters WP, Stahl MT, Murcko M a. Virtual screening—an overview. *Drug Discov Today* 1998;3:160–78.

[6] Doman TN, McGovern SL, Witherbee BJ, Kasten TP, Kurumbail R, Stallings WC, et al. Molecular Docking and High-Throughput Screening for Novel Inhibitors of Protein Tyrosine Phosphatase-1B. *J Med Chem* 2002;45:2213–21.

[7] Durrant JD, McCammon JA. Computer-aided drug-discovery techniques that account for receptor flexibility. *Curr Opin Pharmacol* 2010;10:770–4.

[8] Richards WG. Computer-aided drug design. *Pure Appl Chem* 1994;66:1589–96.

[9] Kalyaanamoorthy S, Chen Y-PP. Structure-based drug design to augment hit discovery. *Drug Discov Today* 2011;16:831–9.

[10] Keseru GM, Makara GM. Hit discovery and hit-to-lead approaches. *Drug Discov Today* 2006;11:741–8.

[11] Sliwoski G, Kothiwale S, Meiler J, Lowe EW. Computational Methods in Drug Discovery. *Pharmacol Rev* 2014;66:334–95.

[12] Zorina A, Stepanchenko N, Novikova G V, Sinetova M, Panichkin VB, Moshkov IE, et al. Eukaryotic-like Ser/Thr protein kinases SpkC/F/K are involved in phosphorylation of GroES in the Cyanobacterium synechocystis. *DNA Res* 2011;18:137–51.

[13] Martin J, Anamika K, Srinivasan N. Classification of protein kinases on the basis of both kinase and non-kinase regions. *PLoS One* 2010;5:e12460.

[14] Arencibia JM, Pastor-Flores D, Bauer AF, Schulze JO, Biondi RM. AGC protein kinases: from structural mechanism of regulation to allosteric drug development for the treatment of human diseases. *Biochim Biophys Acta* 2013;1834:1302–21.

[15] Cohen P, Alessi DR. Kinase drug discovery--what's next in the field? *ACS Chem Biol* 2013;8:96–104.

[16] Chen L, K. Morrow J, T. Tran H, S. Phatak S, Du-Cuny L, Zhang S. From Laptop to Benchtop to Bedside: Structure-based Drug Design on Protein Targets. *Curr Drug Metab* 2012;18:1217–39.

[17] Sotriffer C, Klebe G. Identification and mapping of small-molecule binding sites in proteins: computational tools for structure-based drug design. *Farmaco* 2002;57:243–51.

[18] Levitt DG, Banaszak LJ. POCKET: A computer graphics method for identifying and displaying protein cavities and their surrounding amino acids. *J Mol Graph* 1992;10:229–34.

[19] Hendlich M, Rippmann F, Barnickel G. LIGSITE: Automatic and efficient detection of potential small molecule-binding sites in proteins. *J Mol Graph Model* 1997;15:359–63.

[20] Huang B, Schroeder M. LIGSITEcsc: predicting ligand binding sites using the Connolly surface and degree of conservation. *BMC Struct Biol* 2006;6:19.

[21] Laskowski RA. SURFNET: A program for visualizing molecular surfaces, cavities, and intermolecular interactions. *J Mol Graph* 1995;13:323–30.

[22] Capra JA, Laskowski RA, Thornton JM, Singh M, Funkhouser TA. Predicting protein ligand binding sites by combining evolutionary sequence conservation and 3D structure. *PLoS Comput Biol* 2009;5.

[23] Peters KP, Fauck J, Frömmel C. The automatic search for ligand binding sites in proteins of known three-dimensional structure using only geometric criteria. *J Mol Biol* 1996;256:201–13.

[24] Tan KP, Varadarajan R, Madhusudhan MS. DEPTH: a web server to compute depth and predict small-molecule binding cavities in proteins. *Nucleic Acids Res* 2011;39:W242–W248.

[25] Laurie ATR, Jackson RM. Q-SiteFinder: an energy-based method for the prediction of protein-ligand binding sites. *Bioinformatics* 2005;21:1908–16.

[26] Hernandez M, Ghersi D, Sanchez R. SITEHOUND-web: a server for ligand binding site identification in protein structures. *Nucleic Acids Res* 2009;37:W413–6.

[27] Weisel M, Proschak E, Schneider G. PocketPicker: analysis of ligand binding-sites with shape descriptors. *Chem Cent* J 2007;1:7.

[28] Morita M, Nakamura S, Shimizu K. Highly accurate method for ligand-binding site prediction in unbound state (apo) protein structures. *Proteins* 2008;73:468–79.

[29] Henrich S, Salo-Ahen OMH, Huang B, Rippmann FF, Cruciani G, Wade RC. Computational approaches to identifying and characterizing protein binding sites for ligand design. *J Mol Recognit* 2009;23:209–19.

[30] Laurie ATR, Jackson RM. Methods for the prediction of protein-ligand binding sites for structure-based drug design and virtual ligand screening. *Curr Protein Pept Sci* 2006;7:395–406.

[31] Porter CT, Bartlett GJ, Thornton JM. The Catalytic Site Atlas: a resource of catalytic sites and residues identified in enzymes using structural data. *Nucleic Acids Res* 2004;32:D129–33.

[32] Arakaki AK, Zhang Y, Skolnick J. Large-scale assessment of the utility of low-resolution protein structures for biochemical function assignment. *Bioinformatics* 2004;20:1087–96.

[33] Chikhi R, Sael L, Kihara D. Real-time ligand binding pocket database search using local surface descriptors. *Proteins* 2010;78:2007–28.

[34] Sael L, Kihara D. Detecting local ligand-binding site similarity in nonhomologous proteins by surface patch comparison. *Proteins* 2012;80:1177–95.

[35] Bianchi V, Gherardini PF, Helmer-Citterich M, Ausiello G. Identification of binding pockets in protein structures using a knowledge-based potential derived from local structural similarities. BMC *Bioinformatics 2012;13 Suppl 4:S17.*

[36] Muegge I, Oloff S. Advances in virtual screening. *Drug Discov Today Technol* 2006;3:405–11.

[37] Yuriev E. Challenges and advances in structure-based virtual screening. *Future Med Chem* 2014;6:5–7.

[38] Sousa SF, Cerqueira NMFS a, Fernandes P a, Ramos MJ. Virtual screening in drug design and development. *Comb Chem High Throughput Screen* 2010;13:442–53.

[39] Wermuth CG, Ganellin CR, Lindberg P, Mitscher LA. Glossary of terms used in medicinal chemistry. *Pure Appl Chem* 1998;70:1129–43.

[40] Yang S-Y. Pharmacophore modeling and applications in drug discovery: challenges and recent advances. *Drug Discov Today* 2010;15:444–50.

[41] Zou J, Xie HZ, Yang SY, Chen JJ, Ren JX, Wei YQ. Towards more accurate pharmacophore modeling: Multicomplex-based comprehensive pharmacophore map and most-frequent-feature pharmacophore model of CDK2. *J Mol Graph Model* 2008;27:430–8.

[42] Wolber G, Langer T. LigandScout: 3-D pharmacophores derived from protein-bound ligands and their use as virtual screening filters. *J Chem Inf Model* 2005;45:160–9.

[43] Chen J, Lai L. Pocket v.2: further developments on receptor-based pharmacophore modeling. *J Chem Inf Model* 2006;46:2684–91.

[44] Barillari C, Marcou G, Rognan D. Hot-spots-guided receptor-based pharmacophores (HS-Pharm): a knowledge-based approach to identify ligand-anchoring atoms in protein cavities and prioritize structure-based pharmacophores. *J Chem Inf Model* 2008;48:1396–410.

[45] Tintori C, Corradi V, Magnani M, Manetti F, Botta M. Targets looking for drugs: a multistep computational protocol for the development of structure-based pharmacophores and their applications for hit discovery. *J Chem Inf Model* 2008;48:2166–79.

[46] Sousa SF, Fernandes PA, Ramos MJ. Protein-ligand docking: current status and future challenges. *Proteins* 2006;65:15–26.

[47] Kitchen DB, Decornez H, Furr JR, Bajorath J. Docking and scoring in virtual screening for drug discovery: methods and applications. *Nat Rev Drug Discov* 2004;3:935–49.

[48] Waszkowycz B, Perkins TDJ, Sykes RA, Li J. Large-scale virtual screening for discovering leads in the postgenomic era. *IBM Syst J* 2001;40:360–76.

[49] Morris GM, Huey R, Lindstrom W, Sanner MF, Belew RK, Goodsell DS, et al. AutoDock4 and AutoDockTools4: Automated docking with selective receptor flexibility. *J Comput Chem* 2009;30:2785–91.

[50] Trott O, Olson AJ. AutoDock Vina: improving the speed and accuracy of docking with a new scoring function, efficient optimization, and multithreading. *J Comput Chem 2010*;31:455–61.

[51] Jones G, Willett P, Glen RC, Leach AR, Taylor R. Development and validation of a genetic algorithm for flexible docking. *J Mol Biol* 1997;267:727–48.

[52] Carlson H a. Protein flexibility and drug design: how to hit a moving target. *Curr Opin Chem Biol* 2002;6:447–52.

[53] Cerqueira NMFS a, Bras NF, Fernandes P a, Ramos MJ. MADAMM: a multistaged docking with an automated molecular modeling protocol. *Proteins* 2009;74:192–206.

[54] Cornell WD, Cieplak P, Bayly CI, Gould IR, Merz KM, Ferguson DM, et al. A Second Generation Force Field for the Simulation of Proteins, Nucleic Acids, and Organic Molecules. *J Am Chem Soc* 1995;117:5179–97.

[55] Durrant JD, McCammon JA. Molecular dynamics simulations and drug discovery. *BMC Biol* 2011;9:71.

[56] Yang J, Wang Y, Chen Y. GPU accelerated molecular dynamics simulation of thermal conductivities. *J Comput Phys* 2007;221:799–804.

[57] Liu W, Schmidt B, Voss G, Müller-Wittig W. Accelerating molecular dynamics simulations using Graphics Processing Units with CUDA. *Comput Phys Commun* 2008;179:634–41.

[58] Shan Y, Kim ET, Eastwood MP, Dror RO, Seeliger MA, Shaw DE. How does a drug molecule find its target binding site? *J Am Chem Soc* 2011;133:9181–3.

[59] Shaw DE, Maragakis P, Lindorff-Larsen K, Piana S, Dror RO, Eastwood MP, et al. Atomic-level characterization of the structural dynamics of proteins. *Science* 2010;330:341–6.

[60] Case DA, Cheatham TE, Darden T, Gohlke H, Luo R, Merz KM, et al. The Amber biomolecular simulation programs. *J Comput Chem* 2005;26:1668–88.

[61] Wang J, Wolf RM, Caldwell JW, Kollman PA, Case DA. Development and testing of a general amber force field. *J Comput Chem* 2004;25:1157–74.

[62] Brooks B, Bruccoleri R. CHARMM: A program for macromolecular energy, minimization, and dynamics calculations. *J Comput Chem* 1983;4:187–217.

[63] Kalé L, Skeel R, Bhandarkar M, Brunner R, Gursoy A, Krawetz N, et al. NAMD2: Greater Scalability for Parallel Molecular Dynamics. *J. Comput Phys* 1999;151:283–312.

[64] Phillips JC, Braun R, Wang W, Gumbart J, Tajkhorshid E, Villa E, et al. Scalable molecular dynamics with NAMD. *J Comput Chem* 2005;26:1781–802.

[65] Schames JR, Henchman RH, Siegel JS, Sotriffer C a, Ni H, McCammon JA. Discovery of a novel binding trench in HIV integrase. *J Med Chem* 2004;47:1879–81.

[66] Yarmoluk SM, Nyporko a. Y, Bdzhola VG. Rational design of protein kinase inhibitors. *Biopolym Cell* 2013;29:339–47.

[67] Badrinarayan P, Sastry GN. Rational approaches towards lead optimization of kinase inhibitors: the issue of specificity. *Curr Pharm Des* 2013;19:4714–38.

[68] Zou L, Wang M, Shen Y, Liao J, Li A, Wang M. PKIS: computational identification of protein Kinases for experimentally discovered protein Phosphorylation sites. *BMC Bioinformatics* 2013;14:247.

[69] Suo S-B, Qiu J-D, Shi S-P, Chen X, Liang R-P. PSEA: Kinase-specific prediction and analysis of human phosphorylation substrates. *Sci Rep* 2014;4:4524.

[70] Wong Y-H, Lee T-Y, Liang H-K, Huang C-M, Wang T-Y, Yang Y-H, et al. KinasePhos 2.0: a web server for identifying protein kinase-specific phosphorylation sites based on sequences and coupling patterns. *Nucleic Acids Res* 2007;35:W588–94.

[71] Gao J, Thelen JJ, Dunker a K, Xu D. Musite, a tool for global prediction of general and kinase-specific phosphorylation sites. *Mol Cell Proteomics* 2010;9:2586–600.

[72] Xue Y, Ren J, Gao X, Jin C, Wen L, Yao X. GPS 2.0, a tool to predict kinase-specific phosphorylation sites in hierarchy. *Mol Cell Proteomics* 2008;7:1598–608.

[73] Dinkel H, Chica C, Via A, Gould CM, Jensen LJ, Gibson TJ, et al. Phospho.ELM: a database of phosphorylation sites--update 2011. *Nucleic Acids Res* 2011;39:D261–D267.

[74] Singh NN, Ramji DP. Protein kinase CK2, an important regulator of the inflammatory response? *J Mol Med (Berl)* 2008;86:887–97.

[75] Faust RA, Gapany M, Tristani P, Davis A, Adams GL, Ahmed K. Elevated protein kinase CK2 activity in chromatin of head and neck tumors: Association with malignant transformation. *Cancer Lett* 1996;101:31–5.

[76] Rajeswari M, Santhi N, Bhuvaneswari V. Pharmacophore and Virtual Screening of JAK3 inhibitors. *Bioinformation* 2014;10:157–63.

[77] Jasuja H, Chadha N, Kaur M, Silakari O. Dual inhibitors of Janus kinase 2 and 3 (JAK2/3): designing by pharmacophore- and docking-based virtual screening approach. *Mol Divers* 2014;3:253–67.

[78] Srinivasan P, Chella Perumal P, Sudha a. Discovery of novel inhibitors for Nek6 protein through homology model assisted structure based virtual screening and molecular docking approaches. *Scientific World Journal* 2014;2014:967873.

[79] Araújo PMM, Pinto da Silva L, Esteves da Silva JCG. Comparative theoretical study of the binding of potential cancer-treatment drugs to Checkpoint kinase 1. *Chem Phys Lett* 2014;591:273–6.

[80] Huang X, Cheng CC, Fischmann TO, Duca JS, Richards M, Tadikonda PK, et al. Structure-based design and optimization of 2-aminothiazole-4-carboxamide as a new class of CHK1 inhibitors. *Bioorg Med Chem Lett* 2013;23:2590–4.

[81] Moreira IS, Fernandes PA, Ramos MJ. Accuracy of the numerical solution of the Poisson-Boltzmann equation. *J. Mol. Struct. THEOCHEM,* vol. 729, 2005, p. 11–8.

In: Computational Chemistry
Editor: Daria Bove
ISBN: 978-1-63321-354-8

Chapter 2

COMPUTATIONAL CHEMISTRY INVESTIGATION OF UV FILTERS

Paulo J. O. Ferreira, Luís Pinto da Silva, Margarida S. Miranda and Joaquim C. G. Esteves da Silva*

Centro de Investigação em Química, Departamento de Química e Bioquímica, Faculdade de Ciências da Universidade do Porto, Porto, Portugal

ABSTRACT

The raising level of awareness regarding the harmful effects of solar radiation has resulted in an increase in the production and use of sunscreens. These commercial products contain ultraviolet (UV) filters that absorb, reflect or scatter UV radiation (290-320 nm for UVB and 320-400 nm for UVA) therefore preventing sunburn, photo-aging and ultimately skin diseases such as skin cancer. A high screening efficiency can only be guaranteed if a UV filter has a high photo-stability. However, UV filters are known to undergo degradation upon exposure to sunlight and chemical substances used in water disinfection such as chlorine, leading to the formation of undesirable byproducts with properties different from the parent compounds. This chapter presents a

* Corresponding author: Joaquim C.G. Esteves da Silva, e-mail: jcsilva@fc.up.pt; Tel: (+351) 220-402-569; Fax: (+351) 220-402-659.

comprehensive review on the computational studies performed so far on the structural, electronic, energetic and UV absorption properties of some of the most widely used UV-filters and their degradation products (UVF-DP). *Ab initio* and density functional theory (DFT) methods have been used to obtain the molecular structure, energetics and other molecular properties of UVF-DP. UV-Vis absorption spectra and excited state properties have been calculated using the time dependent density functional theory (TD-DFT). The influence of different solvents in these properties has been studied using solution models. Transformation and degradation mechanisms have been established with the aid of computational methods and results indicate that these can be employed to assess the environmental fate of UV-filters.

1. INTRODUCTION

In the last decades, the increasing knowledge and concern about the harmful effects of the sun`s UV radiation has led to an increase in the production and use of sunscreens. These commercial products contain organic and inorganic molecules known as UV filters that absorb, reflect or scatter UV radiation (290-320 nm for UVB and 320-400 nm for UVA) thus minimizing negative health effects as sunburn, photo-aging and ultimately skin diseases such as skin cancer. Nowadays these compounds are incorporated not only in sunscreens but also into a number of personal care products and even in plastics, varnishes and clothes to enhance their light stability. However, a number of UV filters have shown to present toxic effects, thus maximum concentrations have been established (from 4 to 10%, w/w, according to the EU legislation) with a compromise between adequate protection and minimum side effects for users.

UV filters can be classified into two types: inorganic (also regarded as physical) UV filters, which reflect and scatter radiation and organic (considered chemical) UV filters, which absorb the UV radiation. There are only two approved inorganic UV filters: titanium dioxide and zinc oxide and the organic UV filters comprise various classes of compounds the most common being the cinnamates, benzophenones, dibenzoylmethanes, camphor derivatives, benzimidazoles, *para*-aminobenzoates and salicylates. [1-3] In general these compounds possess one or more benzenic moieties conjugated with electron releasing and electron accepting groups in either *ortho* or *para* positions therefore allowing an efficient electronic delocalization and rendering them a specific maximum absorbance wavelength. They very often also have attached hydrophobic groups like long alkane chains to prevent aqueous dissolution.

UV filters should be stable upon exposure to sunlight. A high screening efficiency can only be guaranteed if a UV filter has a high photostability. However, studies have reported that exposure of UV filters to natural and/or artificial light may lead to photodegradation reactions that can compromise their physical properties (e.g. maximum absorption wavelength and absorbance coefficient) and lead to the formation of undesirable photoproducts that accumulate on human skin. UV filters are also expected to photodegradate after being discharged into the aquatic environment in the presence of sunlight. In addition UV filters can also suffer degradation through reaction with disinfectant products, like chlorine used in swimming pools giving origin to chlorine by-products whose toxic effects are of primary concern. Therefore some attention has been paid in the last years to the degradation of UV filters in aqueous solution. [1,2] The identification of degradation products is important to determine the environmental and human health effects of the use of UV filters as these degradation products normally present more toxic effects than the parent UV filters.

Although the transformation and/or degradation of UV-filters has been primarily studied by using experimental techniques, computational chemistry has been recently applied in parallel with experiment to a better understanding of these processes, once a UV-filter is applied on human skin and subsequently released into the aquatic environment. The increasing availability of powerful computational resources has allowed the evolution of the theoretical methods as well as their application to diverse molecular systems and their chemical and physical properties. It resulted in the possibility of performing high quality investigations for increasingly larger molecular targets, in order to confirm and even extend experimental results. Density functional theory (DFT) is the quantum mechanical method most commonly used in the study of UV-filters and their transformation/degradation processes. It is known to provide an accurate description of the structure, energy, and molecular properties of the ground state of the molecules. It is an important and useful tool that allows a better understanding of molecules and their behavior. It helps to establish the structure-property relationships and also enables the screening of compounds that have the ability to absorb radiation in the desired range. Furthermore, it allows for the prediction and interpretation of the excited-state properties of different types of molecules.

This chapter presents a comprehensive review on the computational studies performed so far on the structural, electronic, energetic and UV absorption properties of some of the most widely used UV-filters and their degradation products (UVF-DP). In section 2.1 we describe cinnamates studies, in section 2.2 camphors studies, in section 2.3 benzophenones studies, in section 2.4 dibenzoylmethanes studies and in section 2.5 benzimidazoles studies.

2. Computational Studies on UVF-DP

2.1. Cinnamates

2-Ethylhexyl-4-methoxycinnamate (EHMC) is one of the most commonly used cinnamates. It presents a high absorption capacity in the UVB region around 310 nm. [3] It is found in commercial sunscreen products in the *trans* isomer ((*E*)-EHMC), however various studies have demonstrated that when exposed to sunlight EHMC suffers isomerization from (*E*) to its (*Z*)-EHMC isomer (see Figure 1). [4] The (*Z*) isomer has a maximum wavelength similar to that of the (*E*) isomer but it has a substantially lower molar absorption coefficient which results in a decreased efficiency of EHMC as a UVB absorber. [4]

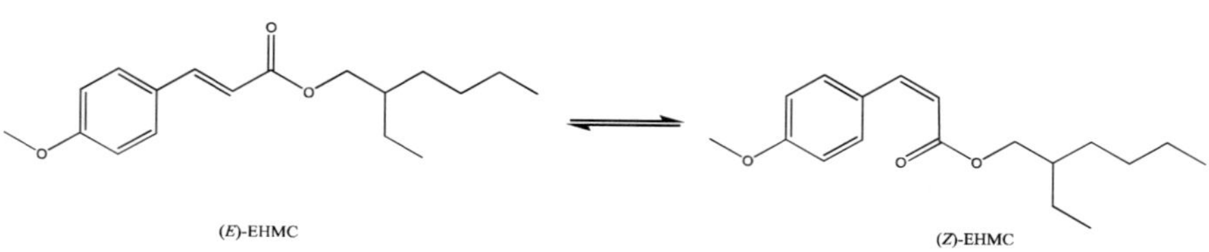

Figure 1. (E) and (Z) isomers of 2-ethylhexyl-4-methoxycinnamate (EHMC).

The photophysical properties of five methoxy-substituted 2-ethylhexylcinnamates including EHMC were studied by Karpkird et al. [5] with experimental (time-resolved fluorescence) and theoretical calculations. Such calculations were performed with semi-empirical methods: PM3 for ground state optimization, and ZINDO/S for estimating the excited state energies and oscillator strengths for both *trans* and *cis* forms of all methoxy-substituted cinnamates. The PM3 calculations demonstrated that all compounds presented planar π-system (*i.e.* C=C and C=O in the same plane as the phenyl ring). The exception was *cis*-2-ethylhexyl-2,4,6-trimethoxycinnamate, for which the C=C and C=O bonds are twisted out of the phenyl ring plane due to steric hindrance by the two *ortho* methoxy groups. These calculations predict that the lowest excited state is of nπ* character for all compounds and that the next two higher states are of ππ* character. Nevertheless, given the estimated radiative rate constants and fluorescence quantum yields, the authors did not believe that the first excited state was not nπ* but a ππ* state. [5] This difference between gas-phase calculations and condensed-phase experiments was already observed, which suggest that more accurate methods than semi-empirical ones are needed for studying the photophysics of UV filters. The same calculations demonstrated that the two ππ* states are more separated in *ortho* and *meta*, while almost degenerate in *trans-para* compounds. Furthermore, it was found that *ortho* and *meta* compounds possess a weaker ππ* state followed by a stronger one, while the contrary was seen for *trans-para*. ZINDO/S//PM3 calculations were also able to provide ππ* states energy differences in line with experimental values. It should be noted that it is this energy difference that dictates the size of the activation energy for non-radiative decay. Thus, these semi-empirical calculations confirmed both the experimentally observed state order for the ππ* states, and the difference in the experimentally determined activation energies for non-radiative decay. The obtained information shows how the performance of the commercial sunscreen agent EHMC might be improved.

In parallel with an experimental study, Alves et al. [6] calculated the ultraviolet absorption spectrum of EHMC using the time dependent density functional theory (TD-DFT) and analyzed the solvent effect using ten different solvents. The conformational search was initially carried out using the PM5 semiempirical method. Several functionals were then employed: B3LYP, PBE0, M06, and PBEPBE with the 6-31+G(2d) basis set for geometry optimizations. The optimized geometries have shown similar results among the methods used. Computational data were compared with experimental data available for methyl *p*-methoxycinnamate (there is no available experimental data for EHMC) and a very good agreement was found. The optimized geometries were used for the TD-DFT calculation of the excited states. The calculations of excited states were carried out using the TD-DFT with the integral equation formalism of the polarizable continuum model (IEF-PCM), into the consistent reaction field, where the solvent is taken into account by means of a polarizable dielectric medium. The absorption maximum absorption wavelength of EHMC was found to be subjected to solvent effects, in which bathochromic shifts were observed in more polar solvents. The TD-PBE1PBE method has shown the best agreement with the experimental results, and that the main energy absorption is due to the HOMO–LUMO transition. The results mainly suggest $\pi \rightarrow \pi^*$ contribution for this transition. M06 and B3LYP were found to show basically the same behavior, while pure functionals were found to have the largest absorption wavelength.

Miranda et al. [7] performed quantum chemical calculations using DFT with the B3LYP density functional and the 6-31G(d) and 6-311++G(d,p) basis sets to study the gas-phase molecular structure of EHMC and its energetic stability. Charge delocalization has been analyzed using natural charges and Wiberg bond indexes within the natural bond orbital analysis and using nucleus independent chemical shifts (NICS). The (*E*) isomer of EHMC was found to be more stable than the (*Z*) by about 20 kJ mol^{-1} in both the gas and aqueous phases. The enthalpy of formation in the gas-phase of (*E*)-EHMC was estimated (-473.5 kJ mol^{-1}) from B3LYP/6-311++G(d,p) calculations using an isodesmic bond separation reaction. Long-range corrected DFT calculations in implicit water were made in order to understand the excited state properties of the (*E*) and (*Z*) isomers of EHMC. The (*E*) and (*Z*) isomers of EHMC were found to present very similar properties at the Franck-Condon state, and contrary to conclusion present in the literature, no weaker light absorption capability was assigned to (*Z*)-EHMC. The lower absorbance capacity of this latter compound in water, in comparison with (*E*)-EHMC, was attributed to its less efficient photo-based formation due to the production of degradation products.

Photodegradation of EHMC in the presence of common oxidizing and chlorinating systems such as H_2O_2, H_2O_2/HCl, H_2O_2/UV and H_2O_2/HCl/UV were recently studied by Gackowska et al. [8] The following chlorinated by-products were identified by GC-MS: chloro-substituted 4-methoxycinnamic acid (4-MCA), 4-methoxybenzaldehyde (4-MBA) and 4-methoxyphenol (4-MP). The experimental work was enriched with MP2 and DFT/B3LYP calculations and the 6-31+G(d,p) basis set. In order to evaluate which chlorinated byproducts are more favorable to form calculations have been performed for all possible mono- and dichloro- isomers of 4-MCA, 4-MBA and 4-MP. In order to include solvent effects the self-consistent reaction field (SCRF) with the polarized continuum model (PCM) and Bondi-type parameterization was used. Global reactivity indices, such as absolute electronegativity, hardness and electrophilicity were calculated. Energy of the lowest unoccupied molecular orbital, E_{LUMO} and the highest occupied molecular orbital, E_{HOMO} were calculated for the optimized at B3LYP/6-31+G(d,p) structures using the MP2/6-31+G(d,p) method. To characterize aromaticity of the studied compounds, a harmonic oscillator model of aromaticity (HOMA) was calculated. The condensed Fukui function was also calculated. Experimental data demonstrated that EHMC photostability decreases in the presence of the reactant oxygen and chlorine species. Spectrophotometric measurements indicated that the reaction proceeds very slowly without UV irradiation. This shows that EHMC is relatively stable in the presence of reactant oxygen and chlorine species when photolysis efficiency is reduced. The DFT thermochemical calculations demonstrated that reactions of 4-MCA, 4-MBA and 4-MP with Cl_2 and HOCl were in all cases thermodynamically favorable. HOCl was found to be a more efficient chlorinating agent than Cl_2. It was found that chlorination reactions do not in all cases lead to the most thermodynamically favorable product. The most stable isomers were not detected experimentally. The thermodynamic analysis did not provide a sufficient explanation of chloroorganic compounds formation. Generally, chlorinated 4-MCA, 4-MBA and 4-MP isomers which are more aromatic and which exhibit higher hardness than others were formed under the experimental conditions. However, in case of monochlorinated 4-MP derivatives, 3-Cl-4-MP, not 2-Cl-4-MP, was detected although 2-Cl-4-MP is slightly more aromatic. Calculations demonstrated that chlorination occurs at the atoms with the highest electron density. The authors found that chlorination reactions occur at the atoms with the highest values of the condensed Fukui function among all sites available for substitution.

2.2. Camphors

4-Methylbenzylidene camphor (4MBC) is a very commonly used UV filter from the camphor derivatives class of UV filters. It absorbs most efficiently at 300 nm and thus is used as an UVB filter. [3] 4MBC can exist as a (*Z*) or (*E*) isomer due to the exocyclic carbon-carbon (styrene) double bond (see Figure 2) and both are chiral. In commercial sunscreen products, 4MBC was shown to consist entirely of (*E*) isomer. [9] However, it is known that, upon exposure to light, (*E*)-4MBC is isomerized reversibly to (*Z*)-4MBC. [10,11] There is no data in the literature regarding the absorption zone of the (*Z*) isomer nor its absorption efficiency but it is known that photoisomerization usually leads to products with different absorption properties, that is a different absorption zone and absorption efficiency.

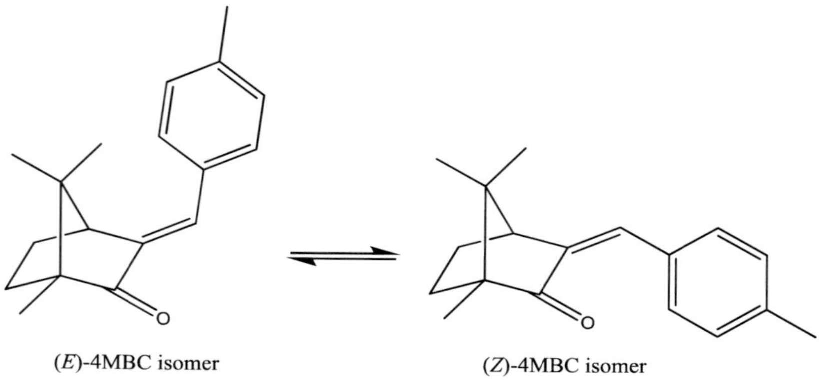

Figure 2. (*E*) and (*Z*) isomers of UVB filter 4-methylbenzylidene camphor (4MBC).

Ferreira et al. [12] performed DFT calculations with the B3LYP functional and two basis sets: 6-31G(d) and 6-311G(d,p) to obtain the gas-phase molecular structure and energetics of the (*E*)- and (*Z*)-isomers of 4MBC. To obtain more accurate energy values the authors also used the G3(MP2)//B3LYP method. The standard molar enthalpy of formation of 4MBC in the gas-phase, at $T = 298.15$ K, was derived from these calculations using appropriately chosen reactions. For comparison and validation purposes these calculations were extended to the parent molecule: camphor. The calculated geometrical parameters for camphor were in good agreement with both crystal and gas phase experimental data as well as *ab initio* calculations (at the HF and MP2 levels) reported in the literature. The authors found the (*E*) isomer was more stable than the (*Z*) by about 10 kJ mol^{-1}. The computational estimates of the gas-phase enthalpy of formation for camphor were in very good agreement with the reported experimental value and this fact supported the estimate obtained for 4MBC -(119.3 ± 0.4) kJ mol^{-1}.

2.3. Benzophenones

Benzophenones (BPs) are widely used in commercial sunscreens due to their high molar absorptivity and ability to absorb in both the UVA and UVB regions of the UV spectrum. [3] Benzophenone-3 (BP-3, 2-hydroxy-4-methoxybenzophenone) is the most commonly used benzophenone.

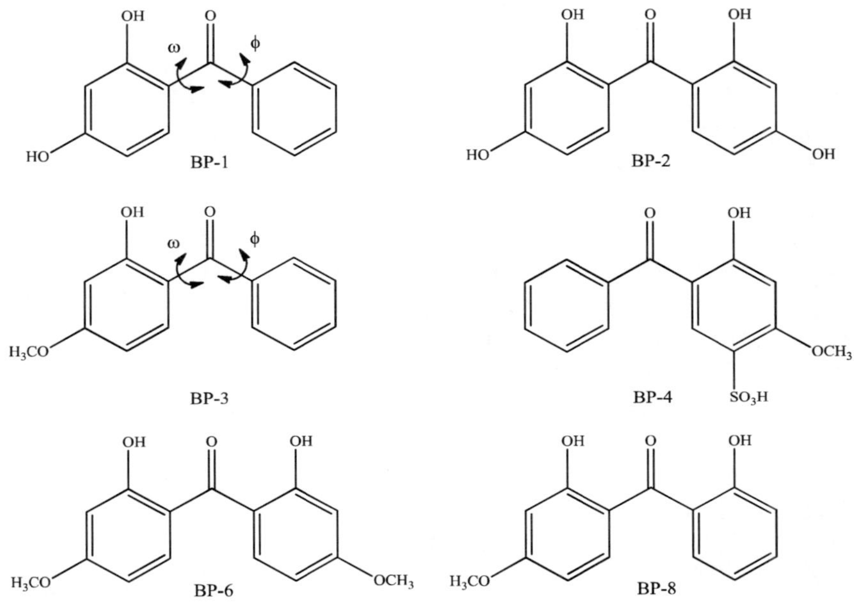

Figure 3. Literature computational studies refer the above benzophenones: (a) benzophenone-1 (BP-1); (b) benzophenone-2 (BP-2); (c) benzophenone-3 (BP-3); (d) benzophenone-6 (BP-6); and (e) benzophenone-8 (BP-8).

Manta et al. [13] carried out a study on the influence of substituents and solvents on the molecular conformation of nine substituted benzophenones including benzophenone-1 (BP-1, 2,4-dihydroxybenzophenone) and BP-3 (see Figure 3). In this study the authors used the HF/6–31G(d) and B3LYP/6–31G(d) methods to find the most stable conformers and the Onsager's and Tomasi's models were used to analyze the solvent effects on the conformers of BPs. The authors expected that this study would be important for the future proposal of relationships between the obtained molecular structural parameters and several experimental physico-chemical properties being determined in the mid time in their laboratory. BP-1 and BP-3 were found to possess non-planar structures as found for the parent benzophenone molecule. The conformers with lower energy were characterized by an intramolecular hydrogen bond (IMHB) between the hydrogen atom of the hydroxyl group at position 2 and the oxygen atom of the carbonyl group (the distances (O···H) are of 1.65-1.66 Å at the B3LYP/6-31G(d) level with Tomasi's method in water, methanol and ethanol) that determines the geometry of these molecules. It was observed that in the gas-phase and water the values of the dihedral angles ω and θ are lower than those exhibited by the parent molecule and ϕ is higher (see Figure 3 for the definition of ω and ϕ; θ corresponds to the inclination angle formed by the planes containing the two aromatic rings). Thus the IMHB increases the planarity of the molecules. The equilibrium constant K_C between the conformer with the IMHB and the one with ω rotated by 180º (without IMHB) have been calculated at the B3LYP/6-31G(d) level. Since K_C<<1 (ranged between 10^{-7} and 10^{-6}) for BP-1 and BP-3 (in the parent benzophenone K_C=1 and there is 50% of each conformer) these BPs are almost completely under the conformation with IMHB in the gas-phase and the studied solvents (the solvents used practically do not affect the constants of BP-1 and BP-3). The dipole of the BPs was related with the solvation parameter *Acity*. It was found that the greater hydrogen bond donating capability of a solvent increases the polarization of the BPs and consequently, increases their dipole moment. The IMHB distances were related with the Hammett`s σ_p constants and it was found that electron donating groups in position 4 such as those present in BP-1 and BP-3 increase the strength of the IMHB.

Lago et al. [14] investigated the thermochemistry and ion energetics of BP-3. This study had to main aims: (1) getting a quantitative estimate of the strength of the IMHB in BP-3; and (2) getting the intrinsic (gas phase) acidity of BP-3 (it is related to the IMHB and, furthermore, allows one to get information on the structure and properties of the anion). The experimental studies were supplemented by quantum chemical calculations at the DFT level. The most stable gas-phase molecular conformation of BP-3 was established through the study of different possible conformers. In general, the computational results for the structure of BP-3 obtained at the B3LYP/6-311++G(d,p) level were in good agreement with the literature available experimental data although some discrepancies relative to the distance between the carbonyl oxygen atom and the hydrogen of the OH group were observed. These small discrepancies can be attributed to the fact that the experiment is performed in crystal, while the calculations refer to the gas phase. The experimentally estimated value for the enthalpy of the IMHB (30.1 ± 6.3) kJ mol^{-1} was found to be in good agreement with the computational estimates 36.8, 35.6 and 28.7 kJ mol^{-1} obtained by means of three isodesmic reaction schemes at the B3LYP/6-311++G(d,p) level. The authors also found that there is a red-shift of the OH stretching vibration in the conformer with an IMHB relative to conformers where that interaction is absent. Computational structural and energetic considerations allowed the authors to classify the hydrogen bond in BP-3 as a “moderate” or “conventional-strong” bond. The calculated theoretical gas phase acidity of BP-3 obtained by taking into account the contributions of all the considered conformers of the molecule and anion: 1395.0 and 1399.3 kJ mol^{-1} at the B3LYP/6-311++G(d,p) and B3LYP/6-311+G(3df,2p) levels, respectively, were found to be close to the experimental value (1402 ± 8.4) kJ mol^{-1}.

Baughman et al. [15] combined experimental and computational techniques to examine the effect that different solvents have on the photophysical properties of three benzophenones commonly used in sunscreens: BP-3, benzophenone-4 (BP-4, 2-hydroxy-4-methoxybenzophenone-5-sulphonic acid) and benzophenone-8 (BP-8, 2,2`-dihydroxy-4-methoxybenzophenone) (see Figure 3). The absorbance spectrum of the three BPs was measured in a wide range of solvents. Computational investigations were performed to examine the experimental results in greater detail and to gain further insight into the molecular-level origin of the unique photophysical properties of three BPs, with a focus on the role that solvent plays in the electronic transitions of the molecules. Geometries were obtained through DFT at the B3PW91/Midi! level and energies were determined through single point B3PW91/6-31+G(d,p) energy calculations. Excitation energies have been calculated using the semi-empirical ZINDO method and the TD-DFT method at the TD-B3PW91/6-31+G(d,p) level. Solvation effects were evaluated using the SCRF Solvent Model. Benzophenone: solvent complexes of the three BPs with a single solvent molecule were investigated in terms of the complexation energies, structural changes due to complexation, and the electronic excitations of the complexed species relative to those of the free chromophore. The energetic profiles of the proposed intramolecular excited state proton transfer (ESPT) reactions were evaluated for all species. For BP-3, the solvent-solute interactions do not have a significant impact upon the observed spectral properties. Calculations support the formation of solvent-solute complexes, but minimal structural changes accompany complexation and excitation energies are not affected. Additionally, the ESPT that is proposed to result in the photostability of the product is found to be thermodynamically favorable and supported by the charge transfer character of the lowest energy (S0 → S1) electronic transition. Similar results are found for BP-4 with the exception of its behavior in electron pair donating solvents. Weakly basic solvents, such as tetrahydrofuran and acetonitrile, were found to be able to solvate the acidic proton to a greater extent resulting in the dominant form of BP-4 as the fully deprotonated anion. In other solvents the acidic proton was found to remain associated with the BP-4. BP-8 showed the greatest solvent dependence in its photophysical behavior. Formation of explicit solvent-solute complexes with both protic and electron pair donating solvents resulted in an increased torsion angle between the ring moieties and an increased excitation energy for the complexed form relative to that of the free species. A complete ESPT is not thermodynamically favorable for BP-8, but examinations of the molecular orbitals and spectral shifts do indicate charge transfer character in the lowest energy electronic transition. The results of this investigation indicate the importance of the solvent-solute interactions upon the photophysical properties and enrich the fundamental understanding of the origin, at a molecular level, of the unique spectral properties of the benzophenone derivatives. This is particularly valuable for the formulation of personal care products utilizing the molecules' absorbance properties, as well as understanding their potential environment impact. Molecular level changes resulting from interactions of the solvent and solute must be considered to fully understand the photophysical properties of these commercially relevant compounds.

Corrêa et al. [16] reported a study of the structural, electronic and spectral properties of twenty five benzophenone derivatives including BP-2, BP-3, benzophenone-6 (BP-6, 2,2`-dihydroxy-4,4`-dimethoxybenzophenone) and BP-8 (see Figure 3), by DFT and TD-DFT methods. Geometries were obtained with the B3LYP density functional and the 6-31G(d) basis set, a level of theory considered to be accurate for optimizing the geometries of benzophenones. [16] The absorption spectra was obtained by means of *in vacuo* TD-B3LYP/6-31G(d) vertical excitation calculations. The lowest ten singlet vertical transitions were computed, in order to obtain their respective excitation energies, oscillator strengths and main transition configurations. Validation of the computational methodology was performed using experimental data from the BP-3: the experimental absorption bands measured at 287 and 325 nm (in dichloromethane) were computed at 285 and 326 nm in the gas-phase. The above mentioned BPs exhibited two peaks one in the UVA and the other in the UVB region. The absorption maxima obtained from the TD-DFT calculations in the gas-phase were in agreement with the experimental absorption bands and showed that the main electronic transitions in the UVA/UVB range present $\pi \rightarrow \pi^*$ character, with the major transition being HOMO → LUMO. These calculations revealed that HOMO, along with other occupied orbitals involved in the excitation, was distributed over the substituted ring. The oscillator strength seems to increase in the presence of disubstitution at the *para* position. For the contrary, the *ortho* position is related to lower oscillator strength. For protic substituents, the position appears to be related to the absorption band. Absorption in the UVB range occurs in the presence of *para* substitution, whereas *ortho* substitution leads to absorption in the UVA spectral region. The obtained results were able to reproduce experimental UV absorption results, and to elucidate the structure-properties relationships of benzophenones. Thus, these results can be useful for customizing absorption properties (wavelengths and intensities) and designing new BP derivatives as sunscreens.

Recently Joseph et al. [17] performed a complete molecular structure and vibrational analyses of BP-3 based on DFT calculations. All possible BP-3 conformers were established from a potential energy surface scan at the B3LYP/6-31G(d) level. The geometry of the most stable conformation was re-optimized at the B3LYP/6-311++G(d,p) level. The calculated molecular structure of BP-3 was found to be non-planar, with the two phenyl rings being inclined at 45.68° with respect to one another. The DFT calculated molecular geometry indicated that the BP-3 molecule exhibits an intra-molecular hydrogen bonding interaction (O-H···O). Detailed vibrational descriptions have been carried out by the authors with the help of normal coordinate analysis. The harmonic-vibrational wavenumbers for BP-3 were calculated at the B3LYP/6-311++G(d,p) level. The difference between the observed and scaled wavenumber values of most of the fundamentals was found to be very small. Therefore, the assignments made at DFT level of theory, with only reasonable deviations from the experimental values, seem to be correct. The red shift of carbonyl and O-H stretching wavenumbers indicated the formation of intra-molecular hydrogen bonding. The various types of intra-molecular electron interactions and their stabilization energies were determined by natural bond orbital analysis. The UV absorption spectrum were examined in ethanol solvent and compared with the calculated one in the gas phase as well as in solvent environment (PCM model) using TD-DFT method employing the B3LYP/6-311++G(d,p) functional/basis set. The strong band observed experimentally at 335 nm was calculated at 325 nm in the gas phase, and at 325 nm in the solvent medium, respectively. Several thermodynamic properties including standard molar heat capacity, standard molar entropy, and standard molar enthalpy changes (from 100 to 500K) were also calculated at the B3LYP/6-311++G(d,p) level.

2.4. Dibenzoylmethanes

Dibenzoylmethane derivatives are widely used as UVA filters and the most common among the dibenzoylmethanes is 4-*tert*-butyl-4`-methoxydibenzoylmethane (BMDBM). In sunscreen products BMDBM exists predominantly in the enol form (see Figure 4) which displays a strong absorption band in the UVA region around 350 nm. [3] Under the presence of sunlight (or artificial irradiation) the *enol* tautomer is known to transform into the *keto* tautomer accounting for the large loss in absorption in the UVA range and thus reducing the effectiveness of the sunscreen.

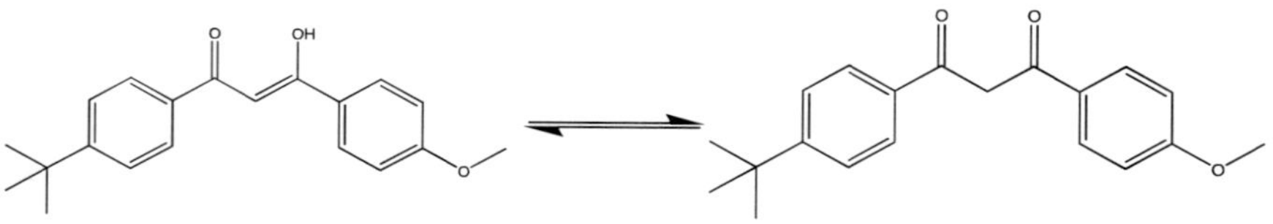

Figure 4. The enol and keto tautomers of 4-tert-butyl-4′-methoxydibenzoylmethane (BMDBM).

The tautomerization processes of *keto* BMDBM were studied photochemically and kinetically by Yamaji et al. [18] The photochemical formation of *keto* BMDBM is shown through the absorption spectral changes during steady-state photolysis of *enol* BMDBM. The photochemical features of the *keto* form were investigated by means of laser flash photolysis. Kinetic studies were also performed into the tautomerization process from *keto* BMDBM to the *enol*. On the basis of the energies of the states of the tautomers and isomers as estimated by DFT calculations, a schematic energy diagram has been derived for the photothermal tautomerization processes of BMDBM in acetonitrile.

Ferreira et al. [19] performed a DFT calculations with the B3LYP density functional and the 6-31G(d) and 6-311++G(d,p) basis sets to obtain the gas-phase molecular structure and energetic stability of the *keto* and the two *enol* tautomers of BMDBM. For comparison and validation purposes a similar study has been performed for the parent molecule dibenzoylmethane (DBM). The *enol* tautomers were found to be more stable than the *keto* due to the existence of an intramolecular resonance-assisted hydrogen bond (RAHB) which was analyzed by QTAIM. The aromatic character of the six-membered *enol* rings was evaluated from NICS calculations. These calculations revealed that the six-membered *enol* ring structure of BMDBM is essentially non-aromatic. Using isodesmic or almost isodesmic reactions the standard molar enthalpy of formation in the gas phase of BMDBM, at T = 298.15 K, was derived. The good agreement between computational estimate and experiment found for DBM gave the authors confidence on the estimate presented for BMDBM (-(372.4 ± 6.4) kJ mol^{-1}).

Pinto da Silva et al. [20] performed DFT calculations with the B3LYP density functional and the 6-31G(d) and 6-311++G(d,p) basis sets to obtain the gas-phase molecular structure and energetic stability of the *enol* and *keto* tautomers of BMDBM. The authors also used the CAM-B3LYP functional in implicit solvent (water, ethanol and cyclohexane) with the 6-31+G(d) basis set in order to analyze the excited-sate properties of the tautomers of BMDBM. The chosen solvents were modeled by using the polarizable continuum implicit model (PCM). The calculations performed correctly describe the excitation wavelengths of the three tautomers of BMDBM, and demonstrated that both *enols* have similar properties at the Franck-Condon state (while the *keto* tautomer presents different properties). The efficiency of ground → excited states transition was rationalized based on the concept of molecular orbital superposition. It was stated that a higher delocalization of HOMO between the dibenzoyl scaffold should increase the excitation efficiency, as the lowest unoccupied orbitals involved in the transitions present such delocalization in all three BMDBM species. It was also found that in general orbital superposition in ethanol and cyclohexane is lower than in water, which is in line with the higher oscillator strength present by the latter solvent.

The loss of UV protection was attributed to the *enol* → *keto* phototautomerism and subsequent C−C bond breaking. This process is not energetically favorable in the singlet bright state, but intersystem crossing to the first two triplet states was found to occur, in which the potential energy surface of C−C bond breaking was more favorable.

2.5. Benzimidazoles

2-Phenylbenzimidazole-5-Sulfonic Acid (PBSA) (see Figure 5) is the most commonly used benzimidazoles due to its strong absorption in the UVB region around 300 nm. [3]

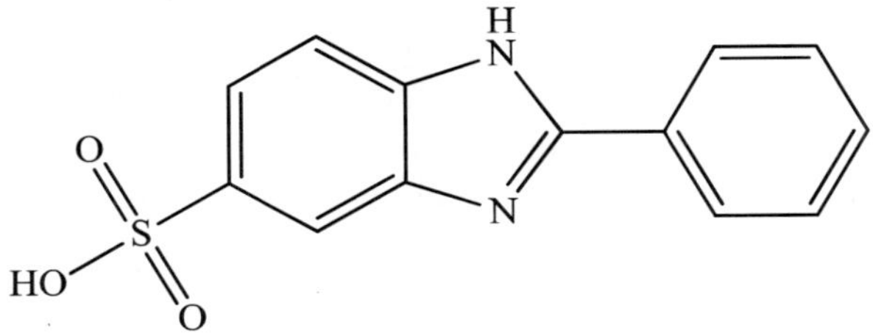

Figure 5. Structure of 2-phenylbenzimidazole-5-sulfonic acid (PBSA).

Zhang et al. [21] developed a computational method based on DFT to predict and evaluate the photodegradation behavior and effects of water constituents on PBSA. Energy and electron transfer reactions of excited state PBSA (PBSA*) with $^3O_2$ and water constituents were evaluated. The computational results indicated that PBSA* could photogenerate $^1O_2$ and $O_2^{-\bullet}$, triplet excited state humic/fulvic acid analogs could not photosensitize the degradation, and the anions (Cl^-, Br^-, and HCO_3^-) could not quench PBSA* or its radical cation chemically. Experiments employing simulated sunlight confirmed that PBSA photodegraded via the direct and self-sensitization mechanism involving $O_2^{-\bullet}$. The photodegradation was shown to be pH-dependent. The direct and self-sensitized photodegradation was found to be inhibited by fulvic acid. The main photodegradation products were identified, and the pathways were clarified. The obtained results indicate that the DFT based computational method can be employed to assess the environmental photochemical fate of organic pollutants.

The kinetics and mechanism of photocatalytic degradation of PBSA in illuminated TiO_2 suspension was studied by Ji et al. [22] Photocatalytic reactions followed pseudo-first-order kinetics. Radical scavenger experiments indicated that $HO^\bullet$ was the predominant radical responsible for an appreciable degradation of PBSA. Second-order rate constant of PBSA-$HO^\bullet$ reaction was determined to be $5.8 \times 10^9\ M^{-1}\ s^{-1}$ by competition kinetics method, suggesting $HO^{\bullet-}$ mediated PBSA oxidation is near the diffusion-controlled limit. Organic intermediates were identified by LC-MS analysis, and the major photoproducts included hydroxylated products, imidazolering cleavage compounds and phenylimidazole carboxylic derivatives. Molecular orbital calculations were carried out at the single determinant (B3LYP/6-311+G(d)) level. The frontier electron densities (FEDs) of the HOMO and the LUMO were determined. Values of FED^2HOMO+ FED^2LUMO were calculated to predict the reaction sites for hydroxyl addition. FEDs calculation predicted that both the two benzenes in PBSA are likely to be attacked by $HO^\bullet$, resulting in the formation of mono-hydroxylation products. Four carboxylic acids, oxalic, malonic, acetic and maleic acids were detected during PBSA photocatalysis. Sulfonic moiety of PBSA was primarily released as a sulfate ion while nitrogen atoms were converted predominantly to ammonium and a less extent to nitrate. Approximately 80% total organic carbon was removed after 720 min irradiation implying mineralization can be expected to obtain after a very long irradiation time. A comparative study on photocatalytic degradation of PBSA and structurally related compounds revealed that the 5-sulfonic moiety had negligible effect on the photocatalysis while 2-phenyl substituent stabilized the benzimidazole ring system to photocatalytic degradation. The study performed by Ji et al. [22] suggests that TiO_2 photocatalysis is a promising treatment technology for sunscreen agent PBSA, a thorough evaluation of the toxicity of intermediates is essential in order to optimize photocatalytic treatment and evaluate the potential risks to the ecology system before the technology is applied for water purification.

Conclusion

In the last decades, the harmful effects of the sun's UV radiation have been demonstrated and divulgated, leading to growing concern by the general population. In response, significant effort was made to develop inorganic and organic molecules that could absorb, reflect or scatter UV radiation. These properties of the so-called UV filters minimize negative health effects as sunburn, photo-aging and ultimately skin diseases. There are only two approved inorganic UV filters, while organic filters comprise various classes of compounds. Thus, organic UV filters are the most studied in order to better UV protecting commercial products.

The experimental study of the diverse classes of organic UV filters revealed very interesting and complex photoinduced processes: photo-isomerization, photo-tautomerism, photo-degradation and intersystem crossing. It was also revealed the solvent- and/or substituent-dependence of compounds, regarding their photophysical properties. Given this, experimental techniques ceased to be the only mean of studying UV filters, and computational chemistry tools have been applied in parallel to obtain a better understanding of these processes.

Computational methods allows us to obtain direct, accurate and detailed information (down to the atomistic level) of the ground and excited state properties of UV filters, thus being an ideal complement to experiment. Moreover, the increasing availability and development of powerful computational resources has allowed the evolution of theoretical methods, as well as their applications to molecular systems of increasing size.

In the present chapter we have reviewed the many contributions of computational chemistry to the study of organic UV filters, in recent years. It can be see that computational methods, particularly density functional theory based ones, can be considered valuables tools in such studies. These methods have been reproducing with success several experimental findings regarding molecular properties of organic UV filters. Moreover, they have been fundamental in elucidating several structure-properties relationships, thereby allowing the customization of absorption properties and the design of new commercial products. Several photoinduced processes were also characterized by computational chemistry, as photo-isomerization and photo-degradation, to a level of detail unreachable by the current experimental techniques. In conclusion, computational methods have gained importance in the study of organic UV filter in recent years, and can now be considered to be valuable and even necessary tools in the research of such compounds.

REFERENCES

[1] Giokas, D. L.; Salvador, A.; Chisvert, A. *Trends Anal. Chem.* 2007, *26*, 360-374.

[2] Díaz-Cruz, M. S.; Llorca, M.; Barceló, D. *Trends Anal. Chem.* 2008, *27*, 873-887.

[3] Shaat, N. A. *Photochem. Photobiol. Sci*. 2010, *9*, 464-469.

[4] Huong, S. P.; Andrieu, V.; Reyner, J.-P.; Rocher, E.; Fourneron, J.-D. *J. Photochem. Photobiol. A: Chem*. 2007, *186*, 65-70.

[5] Karpkird, T. M.; Wanichweacharungruang, S.; Albinsson, B. *Photochem. Photobiol. Sci.* 2009, *8*, 1455-1460.

[6] Alves, L. F.; Gargano, R.; Alcanfor, S. K. B.; Romeiro, L. A. S.; Martins, J. B. L. *Chem. Phys. Lett.* 2011, *516*, 162-165.

[7] Miranda, M. S.; Pinto da Silva, L.; Esteves da Silva, J. C. G. *J. Phys. Org. Chem*. 2014, *27*, 47-56.

[8] Gackowska, A.; Przybyłek, M.; Studziński, W.; Gaca, J. *Cent. Eur. J. Chem.* 2014, *12*, 612-623.

[9] Buser, H.-R.; Muller, M. D.; Balmer, M. E.; Poiger, T.; Buerge, I. *Environ. Sci. Technol*. 2005, *39*, 3013-3019.

[10] Poiger, T.; Buser, H.-R.; Balmer, M. E.; Bergqvist, P.-A.; Muller, M. D. *Chemosphere* 2004, *55*, 951-963.

[11] Rodil, R.; Moeder, M.; Altenburger, R.; Schmitt-Jansen, M. *Anal. Bioanal. Chem*. 2009, *395*, 1513-1524.

[12] Ferreira, P. J. O.; Pinto da Silva, L.; Miranda, M. S.; Esteves da Silva, J. C. G. *Comput. Theor. Chem.* 2014, 1033, 63-73.

[13] Mantas, A.; Blanco, S. E.; Ferretti, F. H. *Internet Electron. J. Mol. Des.* 2004, 3, 387-399.

[14] Lago, A. F.; Jimenez, P.; Herrero, R.; Dávalos, J. Z.; Abboud, J.-L. M. *J. Phys. Chem. A* 2008, *112*, 3201-3208.

[15] Baughman, B. M.; Stennett, E.; Lipner, R. E.; Rudawsky, A. C.; Schmidtke, S. J. *J. Phys. Chem. A* 2009, *113*, 8011-8019.

[16] Corrêa, B. A. M.; Gonçalves, A. S.; de Souza, A. M. T.; Freitas, C. A.; Cabral, L. M.; Albuquerque, M. G.; Castro, H. C.; dos Santos, E. P.; Rodrigues, C. R. *J. Phys. Chem. A* 2012, *116*, 10927-10933.

[17] Joseph, L.; Sajan, D.; Chaitanya, K.; Suthan, T.; Rajesh, N. P.; Isac, J. *Spectrochim. Acta Mol. Biomol. Spectros.* 2014, *120*, 216-227.

[18] Yamaji, M.; Kida, M. *J. Phys. Chem. A* 2013, *117*, 1946-1951.

[19] Ferreira, P. J. O.; Pinto da Silva, L.; Duarte, D. J. R.; Miranda, M. S.; Esteves da Silva, J. C. G. *Comput. Theor. Chem.* 2014, *1038*, 6-16.

[20] Pinto da Silva, L.; Ferreira, P. J. O.; Duarte, D. J. R.; Miranda, M. S.; Esteves da Silva, J. C. G. *J. Phys. Chem. A* 2014, *118*, 1511-1518.

[21] Zhang, S.; Chen, J.; Qiao, X.; Ge, L.; Cai, X.; Na, G. *Environ. Sci. Technol.* 2010, *44*, 7484-7490.

[22] Ji, Y.; Zhou, L.; Ferronato, C.; Salvador, A.; Yang, X.; Chovelon, J.-M. *Appl. Catal. B: Environ.* 2013, *140*, 457-467.

In: Computational Chemistry
Editor: Daria Bove
ISBN: 978-1-63321-354-8

Chapter 3

USING THE NETWORK SIMULATION METHOD TO STUDY IONIC TRANSPORT PROCESSES IN ELECTROCHEMICAL CELLS

A. A. Moya[*]

Universidad de Jaén, Departamento de Física, Jaén, Spain

ABSTRACT

The network simulation method is used to numerically study the steady-state, transient and small-signal ac responses of a binary electrolyte solution placed between two electrodes. The ionic transport processes in the solution are described by the Nernst-Planck and Poisson equations, while the interfacial charge transfer processes are characterized by the Chang-Jaffé equations. The chronoamperometric, chronopotentiometric, linear sweep voltammetric, and small-signal ac responses of symmetric cells, the cation being the only electroactive species at both electrodes, are simulated on the basis of an only network model. The main advantages of the network simulation method with respect to other electrochemical numerical methods, are emphasized. They can be summarized as follows: (i) The network simulation method permits to impose any condition on the electric potential and the electric current; (ii) Steady-state, transient and small-signal ac responses can be dealt with by means of this approach on the basis of an only network

[*] Corresponding author: A. A. Moya. Universidad de Jaén, Departamento de Física, Edificio A-3, Campus Universitario de Las Lagunillas - 23071 Jaén, Spain. E-mail: aamoya@ujaen.es.

model; and (iii) It permits the study of any transport process through spatial regions where large gradients in the ionic concentrations and the electric potential occur.

1. INTRODUCTION

Theoretical study on ionic transport through electrochemical cells is a main topic in Electrochemistry [1]. It is now well known that the ionic transport properties in electrode/solution/electrode systems can be obtained by analysing the response of the system to controlled electric potential or current perturbations. Chronoamperometry, chronopotentiometry, linear sweep voltanmmetry and electrochemical impedance spectroscopy are examples of experimental techniques commonly used to characterizing the electrical properties of simple electrochemical cells [2].

The characterization of electrochemical systems by electrical response measurements requires the development of mathematical models for which the response can be determined theoretically and which serve as a basis for the interpretation of experimental data. For the case of electrode/solution/electrode systems, the theoretical treatment most commonly employed is based on a model in which the ionic transport through the solution is governed by the Nernst-Planck and Poisson equations, while the charge transfer at the interfaces is described by the Chang-Jaffé equations. However, the nature of the Nernst-Planck and Poisson equations is such that an exact analytical solution for the electrical response of electrochemical cells is almost impossible to obtain in the most of the interesting physical situations. Although some of the electrical properties of certain types of electrochemical cells can be analytically obtained by using some approximations, such as the electrical neutrality condition in the solution, infinite charge transfer rate at the electrodes or the Donnan equilibrium relations at the interfaces, the general and rigorous treatment of the problem requires the use of advanced numerical methods.

Excellent procedures have been developed to solve the Nernst-Planck and Poisson equations in electrochemical cells [3-10]. However, they have found serious difficulties in adapting to the different experimental conditions.

An alternative method to the classical numerical methods is the network simulation method [11]. This simulation method basically consists in modelling a physicochemical process by means of a graphical representation analogous to electrical circuit diagrams which is analysed by means of an electric circuit simulation program. Highly developed, commercially available software for circuit analysis can thus be employed to obtain the dynamic behaviour of the whole system without having to deal with the solution of the governing differential equations. In this work, it is shown that the network simulation method can be used satisfactorily in the study of the general electrical response of electrochemical cells.

Some of the reasons for this have been given in our previous papers [12-15], and they can be summarized as follows: (i) The network simulation method permits to impose any condition on the electric potential and the electric current; (ii) Steady-state, transient and small-signal ac responses can be dealt with by means of this approach; and (iii) It permits the study of any transport process through spatial regions where large gradients in the ionic concentrations and the electric potential occur.

Moreover, it must be noted that the network model of the system is an appropriate electric circuit to model the diffusion-migration impedance of electrode/solution/electrode systems [16].

In this paper, a network model for the Nernst-Planck and Poisson equations with boundary conditions of the Chang-Jaffé type is used to describe the ionic transport of a binary electrolyte in a solution placed between two electrodes, being the cation the only electro-active species at both electrodes. With this model and the general purpose electric circuit simulation program PSpice®, the chronoamperometric, linear sweep voltammetric, chronopotentiometric, and small-signal ac responses are simulated.

It will be analysed the influence of the rate constants of the interfacial kinetics on: (i) The steady-state current-voltage characteristic, (ii) The time evolution of the electric current to a pulse and to a linear sweep of electric potential, (iii) The time evolution of the electric potential in response to a pulse of the electric current, and (iv) The electrochemical impedance.

Some of the results obtained can be compared with the analytical solutions in simplified systems or with the results reported in previous studies in this field [17-19].

The results find practical applications in Li-ion or Na-ion electrolytes used in rechargeable batteries [20, 21].

2. Theoretical Description

Let us consider a system constituted by a solution with a 1:1 binary electrolyte placed between two electrodes. Ionic transport process is supposed to be one-dimensional and perpendicular to the electrode|solution interface, with x the direction of transport.

The equations in dimensionless form (Appendix A) determining the behaviour of the system at time t in the electrolyte solution lying from x=-L to x=L, are the laws of mass conservation or continuity equations:

$$\frac{\partial J_i(x,t)}{\partial x} = -\frac{\partial c_i(x,t)}{\partial t}, \qquad i = 1, 2 \tag{1}$$

the Nernst-Planck flux equations written for dilute solutions:

$$J_i(x,t) = -D_{iS}\left[\frac{\partial c_i(x,t)}{\partial x} + z_i\, c_i(x,t)\frac{\partial \phi(x,t)}{\partial x}\right] \tag{2}$$

and the Poisson equation:

$$\frac{\partial \boldsymbol{D}(x,t)}{\partial x} = z_1\, c_1(x,t) + z_2\, c_2(x,t) \tag{3}$$

where:

$$\boldsymbol{D}(x,t) = \varepsilon\, E(x,t) = -\varepsilon\frac{\partial \phi(x,t)}{\partial x} \tag{4}$$

here J_i, D_i, c_i and z_i denote the ionic flux, the diffusion coefficient, the molar concentration and the charge number of ion i (z_1=1 and z_2=-1), respectively. The electric potential is represented by ϕ, electric permittivity by ε, electric displacement by $\boldsymbol{D}$, and electric field by E. The constants F, R, and T have their usual meanings: Faraday constant, ideal gas constant, and absolute temperature, respectively.

On the other hand, the total electric current, I, is the sum of the faradaic and the displacement electric currents, and it is given by:

$$I(x,t) = z_1\, J_1(x,t) + z_2\, J_2(x,t) + \frac{\partial \boldsymbol{D}(x,t)}{\partial x} \tag{5}$$

and from equations 1, 3 and 5, one obtains:

$$\frac{\partial I(x,t)}{\partial x} = 0 \tag{6}$$

i.e., the total electric current is not a function of x, $I(x,t)=I(t)$.

Let us study the response to an externally-applied electric perturbation of an initially electroneutral system characterized by the following equations:

$$c_1(x,0) = c_2(x,0) = c^0 \tag{7a}$$

$$\phi(x,0) = 0 \tag{7b}$$

If we consider the cation to be the only electroactive species and the charge transfer at the electrode/solution interfaces to be described by the Chang-Jaffé equations [4], the boundary conditions for the Nernst-Planck flux equations are:

$$J_1(\pm L,t) = \pm k_{1\pm}\left[c_1(\pm L,t) - c^0\right] \tag{8a}$$

$$J_2(\pm L,t) = 0 \tag{8b}$$

where $k_{1\pm}$ are the kinetic rate constants. The Chang-Jaffé equations involve the physical assumption that the faradaic current arising from the reaction of a charge carrier of a given species is proportional to the excess concentration of that species at the interface. In this paper, ions are assumed to be point charges, which is the basic assumption in the Gouy-Chapman model for the diffuse electrical double layer.

Therefore, the Butler-Volmer equations describing the electrode kinetics reduce to the Chang-Jaffe boundary conditions [5].

In order to study the response of an electrochemical cell to an external electric perturbation, the following conditions must be used:

$$\frac{d\boldsymbol{D}(-L,t)}{dt} = I(t) - z_1 J_1(-L,t) - z_2 J_2(-L,t) \tag{9a}$$

$$\phi(-L,t) = \phi_A(t) \tag{9b}$$

$$\phi(L,t) = 0 \tag{10}$$

Equation 9a, which is obtained from Equation 5, is a boundary condition for the time evolution of the electric displacement, $\boldsymbol{D}$, at x=-L, and it is imposed when the total electric current, $I(t)$, is the externally controlled variable. When the electric potential, $\phi_A(t)$, is the externally applied perturbation, it is necessary to take into account the boundary condition given by Equation 9b, instead of Equation 9a; in this case, the electric current through the system must be evaluated from Equation 5. Finally, Equation 10 defines the reference level for the electric potential.

3. The Network Model

The network model is obtained from a similar viewpoint to that of a finite difference scheme by dividing the physical region of interest, which we consider to have a unit cross-sectional area, into N volume elements or compartments of width δ_k (k=1, ..., N), small enough for the spatial variations of the parameters within each compartment to be negligible [1].

Obviously, the precision of quantitative modeling greatly increases as the degree of subdivision increases too, since in that case we approach the continuum behavior.

The network model for the diffusion-migration process in a compartment of width δ_k, which is extended from x_k-$\delta_k/2$ to x_k+$\delta_k/2$, is shown in Figure 1a, and a complete explanation of the general procedure to obtain it has been given elsewhere [22-24]. In this figure, the network elements are as follows: R_{dk} is the resistor representing the diffusion of ion i in the compartment k; $GJ_{eik}(\pm)$ are the multivariable voltage-controlled current sources modelling the electrical contributions to the flux of ion i, minus and plus meaning respectively the flux entering and leaving compartment k; C_{dk} is the capacitor representing the non-stationary effects of the electrodiffusion process in the compartment k; R_{pk} is the resistor modeling the constitutive equation of the medium; and GJ_{pk} is the voltage-controlled current source modeling the electric charge stores in the compartment k. The relation between those network elements and the parameters of the system is given by:

$$R_{dik} = \frac{\delta_k}{2D_i} \tag{11a}$$

$$GJ_{eik}(\pm) = \pm D_i z_i c_i (x_k \pm \frac{\delta_k}{2}) \frac{\phi(x_k) - \phi(x_k \pm \frac{\delta_k}{2})}{\delta_k / 2} \tag{11b}$$

$$C_{dk} = \delta_k \tag{11c}$$

$$R_{pk} = \frac{\delta_k}{2\varepsilon} \tag{11d}$$

$$GJ_{pk} = -\delta_k \left[z_1 c_1 (x_k) + z_2 c_2 (x_k) \right] \tag{11e}$$

For network modelling purposes, a number N of circuit elements like those in Figure 1a (k=1, ..., N) must be connected in series to form a network model for the entire physical region. Figure 1b shows the network model for an electrochemical cell. In this figure, each of the six-terminals box is constituted by the combination in series of N elements such as those shown in Figure 1a. The next step is to include the initial and boundary conditions of the system in the network model. The initial conditions for concentrations and electric potential (Equation (7a) and Equation (7b)) are included via the initial voltages at the appropriate nodes of the network. The boundary conditions given by Equation (8a) are represented by resistors of resistances $1/k_{\pm}$ connected between the nodes $c_1(\pm L)$ and an independent voltage source of value c^0. The boundary conditions given by Equation (8b) and Equation (10) are introduced respectively by open-circuiting the nodes c_2 at $x=\pm L$ and by short-circuiting the node ϕ at $x=L$. The perturbation of electric current (Equation 9a) is introduced from a current-controlled current source, G_I, which impose the value of the electric displacement at $x=-L$ from the value of the external electric current, I. The perturbation of electric potential (Equation 9b) is represented by an independent voltage source of value $\phi_A(t)$.

4. RESULTS AND DISCUSSION

The results for the transient and small-signal ac electrical properties of electrochemical cells are obtained by means of the simulation into the general purpose electric circuit simulation package PSpice® (from Cadence Design Systems), under transient or ac conditions, of the network model depicted in Figure 1b, with the appropriate numerical values for the system parameters.

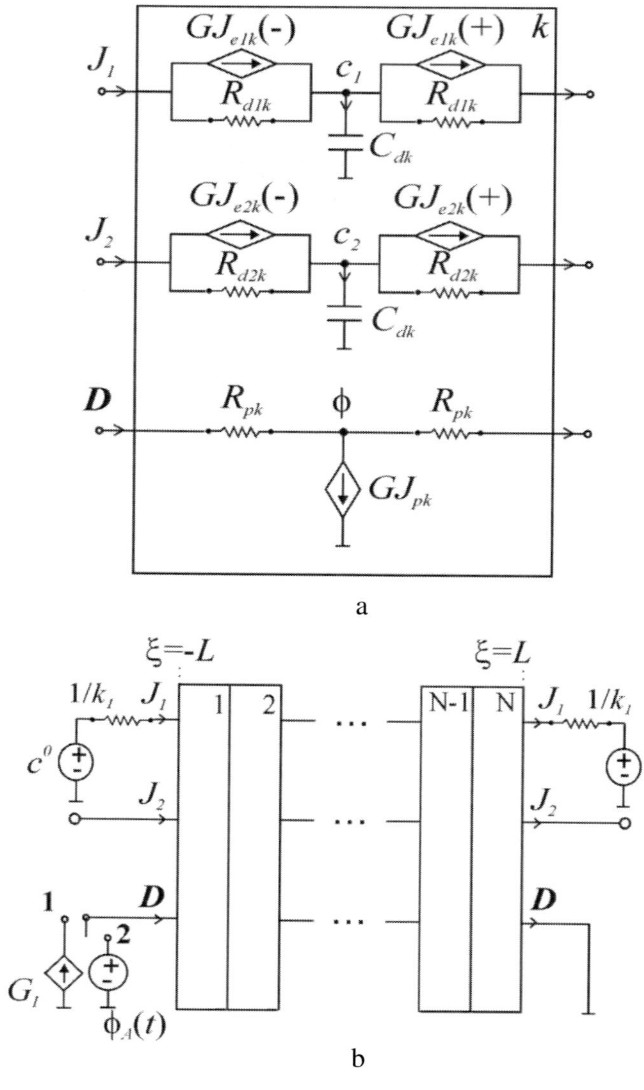

Figure 1. (a) Network model for the diffusion-migration process in a volume element of an electrolyte solution. (b) Network model for an electrochemical cell. Each of the six-terminals box is obtained by the combination in series of N structures such as those shown in (a).

We will now study the electrical responses of symmetric electrochemical cells with various values of the kinetic rate constants $k_{1-}=k_{1+}=k_1$, $z_1=1$, $z_2=-1$, $L=500$, $D_1=2\cdot 10^4$, $D_2=10^4$, $c^0=0.5$ and $\varepsilon=1$. This system could correspond to a LiCl solution placed between two identical electrodes of metal-parent type such as $LiCoO_2$. It must be noted that the limit value $k_1=0$ corresponds to blocking electrodes while the limit $k_1\rightarrow\infty$ correspond to reversible electrodes.

The spatial grid (N and δ_k) chosen considers the presence of two electrode/ solution interfaces and a solution phase. Moreover, it takes into account that the two interfacial regions have a size of about 4λ in each phase. This spatial grid is symmetrical about the cell center. In the right interface, we have chosen 200 compartments of width $(L-10)/100$ from $x=0$ to $x=L-10$, 200 compartments of width 0.03 from $x=L-10$ to $x=L-4$, and 200 compartments of width 0.02 from $x=L-4$ to $x=L$. In this way, $N=1200$ compartments have been used.

4.1. Chronoamperometric Response

In this section, we will report the transient response of an electrochemical cell to an externally applied step of electric potential, $\phi_A(\tau)$. This electric signal is given by:

$$\phi_A(t) = \begin{cases} 0, & t=0 \\ \phi_S, & t>0 \end{cases} \tag{12}$$

where ϕ_S is the amplitude of the perturbing electric potential.

Figure 2 gives the steady-state current-voltage characteristic of the cell for various values of kinetic rate constant, namely $k_1=20$, 50, 100, and ∞. It is worth noting that, when the applied electric potential is positive, $\phi_S>0$, the electric current is also positive, $I>0$, and the cations move from the left to the right interface. In Figure 2, it is observed an ohmic region for small values of ϕ_A (underlimiting current regime) and one region in which the steady-state electric current, I_{SS}, varies very slowly with the voltage (overlimiting current regimes) for high values of ϕ_S. In the lower voltage range, the slope of the steady-state current-voltage curves increases as the rate constant, k_1, increases. In the highest voltage region, the value of the limiting current increases as k_1 increases. When the kinetic rate constants are slow enough, the limiting current is determined by the interfacial rates and occurs when the surface concentration of the positive ion goes to zero at the left interface, i.e.:

$$I_L = k_1 c^0 \tag{13a}$$

For the fastest interfacial reactions, the limiting current is determined by the depletion of ions in the solution phase close to the left interface.

In an electroneutral system, it is given by [19]:

$$I_L = \frac{2 D_1 c^0}{L} \tag{13b}$$

In this way, the threshold value of the kinetic rate constant, k_1^*, given by:

$$k_1^* = \frac{2 D_1}{L} \tag{13c}$$

determines the value of the limiting current. The results obtained from the numerical simulation are then in good agreement when those theoretically expected. The threshold value of the rate constant is $k_1^* = 40$, and the theoretical limiting currents are I_L=10 for k_1=20, I_L=25 for k_1=50, and I_L=40 for $k_1 \geq 40$.

Figure 3 gives the total electric current, $I(t)$, for the amplitudes of the perturbing electric potential ϕ_S=1, 2 and 4, in an electrochemical cell with the kinetic rate constant k_1=20. The initial state is an equilibrium one while the final state is not, since in this final state a steady-state electric current, I_{SS}, passes through the cell. The current-time responses show an initial ohmic current region, followed by a decay into the steady-state current region. The initial values for the electric current are in excellent agreement with those theoretically expected and given by the following equation:

$$I(t = 0^+) = \frac{(D_1 + D_2) c^0}{2L} \phi_S \tag{14}$$

In this way, the initial value of the electric current is directly proportional to the amplitude of the electric potential perturbation: 15 for ϕ_S=1, 30 for ϕ_S=2, and 60 for ϕ_S=4.

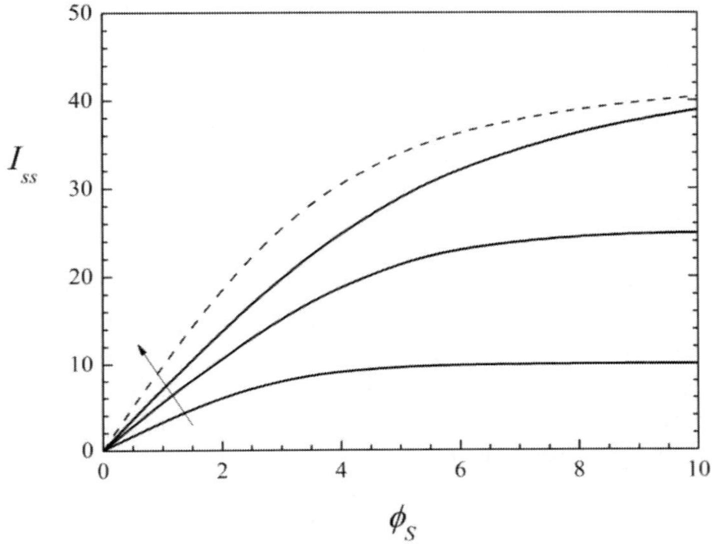

Figure 2. Steady-state current-voltage characteristic for a cell with the values of the kinetic rate constant k_l=10, 20, 50, and ∞. The arrow indicates increasing values of k_l.

However, this behaviour is not observed for the steady-state values of the electric current because of the existence of a limiting current. It is interesting to note that the diffusional relaxation time of cell, t_D, is:

$$t_D = \frac{(2L)^2}{D_2} \tag{15}$$

and for the systems here considered ($2L=10^3$ and $D_2=10^4$) one obtains $\tau_D=100$.

4.2. Linear Sweep Voltammetry

The linear sweep voltammetric response of an electrochemical cell is obtained by perturbing the system with an externally applied electric potential, $\phi_A(\tau)$, given by:

$$\phi_A(t) = vt \tag{16a}$$

here, v is the sweep rate, which can be expresses as:

$$v = \frac{\phi_R}{t_R} \tag{16b}$$

where ϕ_R and t_R are reference values for the electric potential and time, respectively.

Figure 4 gives the current-voltage characteristics of a cell with the kinetic rate constant k_l=50, corresponding to a linear sweep voltammetry with t_R=100 and different values of the reference electric potential, namely, ϕ_S=10, 20, and 40, i.e., v=0.1, 0.2, and 0.4. For the lowest sweep rates, it is possible to obtain the steady-state current-voltage characteristic, as it is usually done in the labs. However, when the sweep rate increases, a peak appears in the current-voltage characteristic. The value of the electric current linearly increases with the electric potential, it then reaches a maximum or peak value, and finally decays to the limiting value. In Figure 4 it is observed that the value of the maximum of the electric current increases as the sweep rate, v, increases.

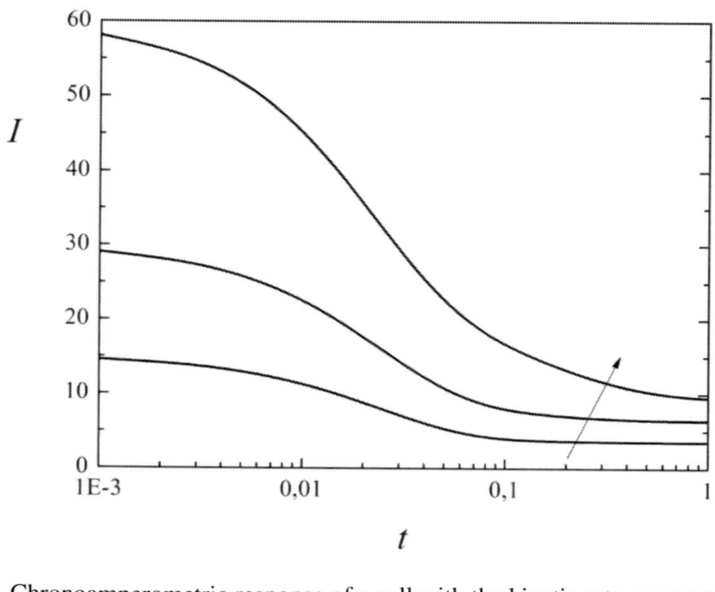

Figure 3. Chronoamperometric response of a cell with the kinetic rate constant k_l=20, for the amplitudes of the perturbing electric potential ϕ_S=1, 2 and 4. The arrow indicates increasing values of ϕ_S.

This behaviour is similar to that found in systems with transport processes controlled by simple diffusion [2].

4.3. Chronopotentiometric Response

In this section, we will report the transient response of a symmetric electrochemical cell to an externally applied electric current, $I(\tau)$. Now, the system is perturbed with an electric signal given by:

$$I(t) = \begin{cases} 0, & t=0 \\ I_S, & t>0 \end{cases} \tag{17}$$

where I_S is the amplitude of the perturbing electric potential.

The time evolution of the electric potential difference across a cell, $\phi_C(t)=\phi(-L,t)$, with reversible electrodes, $k_I \to \infty$, for three different values of the amplitude of the electric current in the underlimiting regime, namely, I_S=10, 20, and 30, are shown in Figure 5.

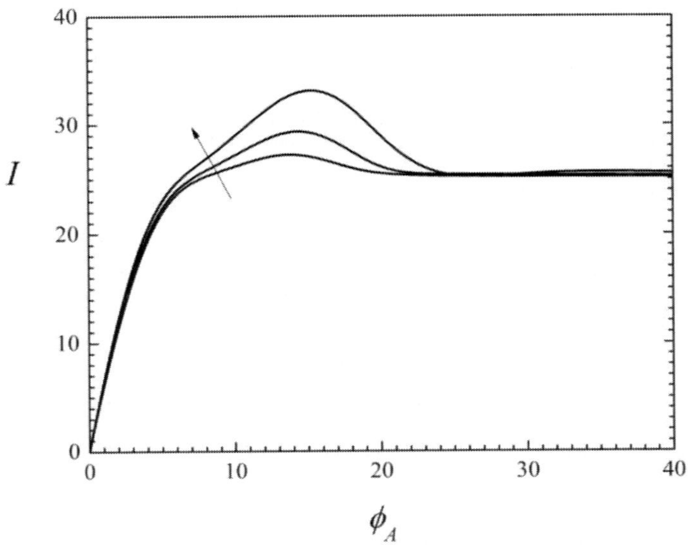

Figure 4. Current-voltage characteristics of a cell with k_I=50, corresponding to a linear sweep voltammetry with the sweep rates v=0.1, 0.2, and 0.4. The arrow indicates increasing values of v.

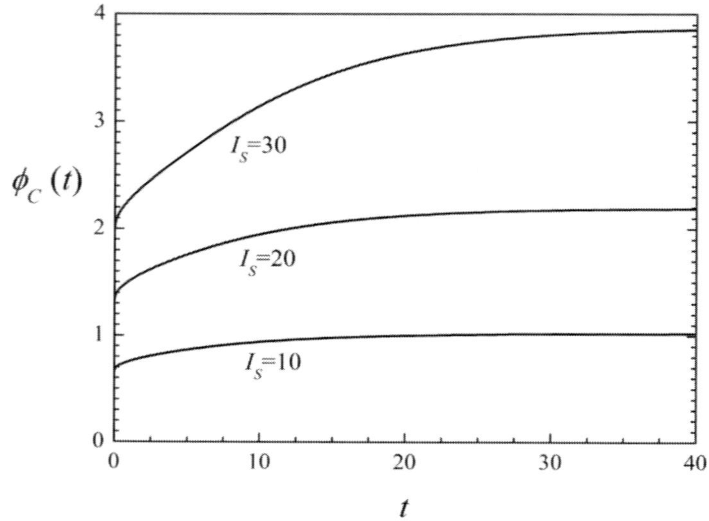

Figure 5. Chronopotentiometric response of a cell with the kinetic rate constant $k_I \to \infty$, for the amplitudes of the perturbing electric current I_S=10, 20 and 30.

In this figure, it can be observed that, after an electric potential vertical jump due to the ohmic resistance of the system, the electric potential difference increases and then it evolves to the steady-state value.

The initial values for the electric potential are in good agreement with those theoretically expected and given by the following equation:

$$\phi_C\,(t=0^+) \;=\; \frac{2L}{(D_1+D_2)\,c^0}\,I_S \tag{18}$$

In this way, the initial value of the electric potential is directly proportional to the amplitude of the electric current perturbation, 0.667 for I_S=10, 1.33 for I_S=20, and 2 for I_S=30. However, this behaviour is not observed for the steady-state values of the electric current because the existence of a limiting current.

4.4. Electrochemical Impedance

In order to study the electrochemical impedance, the system is perturbed around an equilibrium state with an electric current given by:

$$I(t) = I_0 \sin(\omega t) \tag{19}$$

where I_0 is the amplitude of the perturbing sine electric current and ω is the angular frequency, which can be written as a function of the conventional frequency, f:

$$\omega = 2\pi f \tag{20}$$

When a small-amplitude ac signal is used, the electric potential of the cell can be written as:

$$\phi_C(t) = \phi_0 \sin(\omega t + \varphi) \tag{21}$$

where ϕ_0 is the ac amplitude of the electric potential of the cell and φ is the phase difference between the voltage and the perturbing current. Now, the electrochemical impedance, Z, is a complex quantity given by the following equation [26]:

$$Z(j\omega) = \frac{\phi_0}{I_0} e^{j\varphi} = Z_r(\omega) + j Z_i(\omega) \tag{22}$$

where $j = \sqrt{-1}$ is the imaginary unit, and Z_r and Z_i are the real and imaginary parts of the impedance, respectively.

Figure 6 gives the complex-plane impedance plot of an electrochemical cell with the kinetic rate constant k_1=50. In this plot, $-Z_i(\omega)$ is plotted against $Z_r(\omega)$ with the angular frequency, ω, as a parameter increasing from the right to the left of the plot. This Nyquist plot shows three regions associated with single processes [23]: a geometric semicircle at high frequencies, an interfacial semicircle at intermediate frequencies, and the finite-length Warburg diffusion impedance at low frequencies. The characteristic frequencies of the geometric, interfacial and diffusion arcs respectively are $2.4 \cdot 10^3$, 3.9, and 0.022, and they are in good agreement with the theoretical estimations. The characteristic frequency of the high-frequency geometric arc is given by:

$$f_g = \frac{(D_1 + D_2)c^0}{2\pi\varepsilon} \tag{23a}$$

The characteristic frequency of the intermediate-frequency interfacial arc is:

$$f_i = \frac{k_1 c^0}{2\pi\sqrt{2\varepsilon c^0}} \tag{23b}$$

and the characteristic frequency of the low-frequency diffusion arc is [27]:

$$f_d = \frac{2.54}{2\pi} \frac{2 D_1 D_2}{(D_1 + D_2) L^2} \tag{23c}$$

For the systems here considered one obtains f_g=2.4·10^3, f_i=3.98 and f_d=0.022.

On the other hand, the dc resistance of the system, i.e., the value of the real part of the impedance at the limit of zero frequency is 0.18, being this value in excellent agreement with that theoretically expected:

$$R_{DC} = \frac{2L}{D_1 c^0} + \frac{2}{k_1 c^0} \tag{24}$$

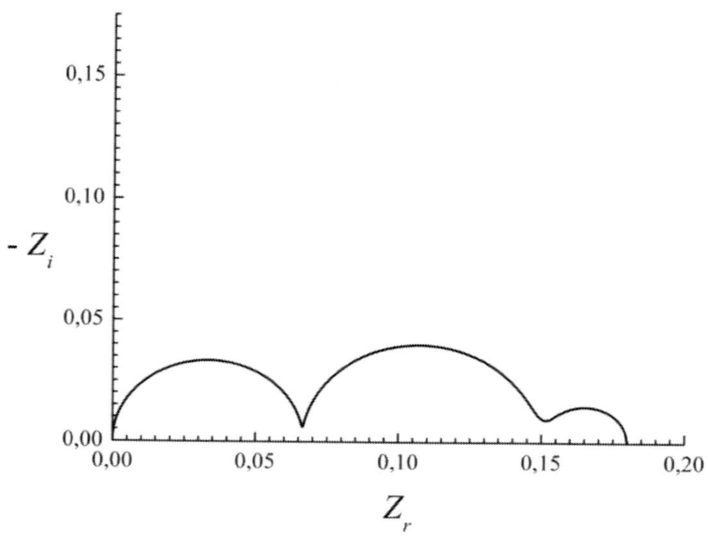

Figure 6. Complex-plane impedance plot for an electrochemical cell with the kinetic rate constant k_1=50.

CONCLUSION

It has been shown that the network simulation method allows us to numerically study the steady-state, transient and small-signal ac responses of a binary electrolyte solution placed between two electrodes. The ionic transport processes in the solution are described by the Nernst-Planck and Poisson equations, while the interfacial charge transfer processes are characterized by the Chang-Jaffé equations. The chronoamperometric, chronopotentiometric, linear sweep voltammetric, and small-signal ac responses of symmetric cells, the cation being the only electroactive species at both electrodes, are simulated on the basis of an only network model. The main advantages of the network simulation method with respect to other electrochemical numerical methods can be summarized as follows: (i) The network simulation method permits to impose any condition on the electric potential and the electric current; (ii) Steady-state, transient and small-signal ac responses can be dealt with by means of this approach on the basis an only network model; and (iii) It permits the study of any transport process through spatial regions where large gradients in the ionic concentrations and the electric potential occur.

APPENDIX A

In this chapter, the study is presented by using dimensionless variables. They are obtained by dividing the variable by the following scaling factors:

- Molar concentration and fixed-charge concentration (mol m^{-3}): c_a
- Diffusion coefficient (m^2 s^{-1}): D_a
- Position and length (m): λ
- Time (s): $\dfrac{\lambda^2}{D_a}$
- Ionic flux (mol m^{-2} s^{-1}): $\dfrac{D_a\, c_a}{\lambda}$
- Electric potential (V): $\dfrac{RT}{F}$

- Electric field ($V\ m^{-1}$): $\dfrac{RT}{F\lambda}$

- Electric current ($A\ m^{-2}$): $\dfrac{F\,D_a\,c_a}{\lambda}$

- Resistance ($\Omega\ m^2$): $\dfrac{RT\lambda}{F^2\,D_a\,c_a}$

- Frequency (s^{-1}): $\dfrac{D_a}{\lambda^2}$

here, λ, D_a and c_a are scaling factors with the dimensions of length, diffusion coefficient and molar concentration, respectively. These three variables are chosen as characteristic values of the system studied. In particular, if one denotes by ε the electric permittivity of the medium ($C\ V^{-1}\ m^{-1}$), the length λ given by:

$$\lambda = \sqrt{\frac{\varepsilon RT}{F^2\,c_a}}$$

can be considered as the Debye length in the system. On the other hand, the constants F, R and T have their usual meanings: Faraday constant ($C\ mol^{-1}$), ideal gas constant ($J\ K^{-1}\ mol^{-1}$) and absolute temperature (K), respectively.

Using typical values of the diffusion coefficient and the ionic concentration, which lead to $D_a=10^{-11}\ m^2s^{-1}$ and $c_a=20$ mM, respectively, one obtains that the Debye length of the system, λ, is approximately 3 nm. Thus, 1 unit of frequency, electric potential, electric current and resistance is, respectively, 1 MHz, 25 mV, 0.6 $A\ cm^{-2}$ and 0.04 $\Omega\ cm^2$.

REFERENCES

[1] Bockris, J. O'M., Reddy, A. K. N. *Modern Electrochemistry 1*, second edition; Plenum Press: New York and London, 1998.

[2] Bard, A. J., Faulkner, L. R., *Electrochemical methods, fundamental and applications*; Wiley: New York, 1980.

[3] Franceschetti, D. R., Macdonald, J. R. *J. Electroanal. Chem.* 1979, 100, 583.
[4] Brumleve, T. R., Buck, R. P. *J. Electroanal. Chem.* 1978, 90, 1.
[5] Murphy, W. D., Manzanares, J. A., Mafe, S., Reiss, H. *J. Phys. Chem.* 1992, 96, 9983.
[6] Freire, F. C. M., Barbero, G., Scalerandi, M. *Phys. Rev. E* 2006, 73, 051 202.
[7] Beunis, F., Strubbe, F., Marescaux, M., Beeckman, J., Neyts, K., Verschueren, A. R. M. *Phys. Rev. E* 2008, 78, 011502.
[8] Van Soestbergen, M. *Electrochim. Acta* 2010, 35, 1848.
[9] Dickinson, E. J. F., Freitag, L., Compton, R. G. *J. Phys. Chem. B* 2010, 114, 187.
[10] Dickinson, E. J. F., Ekstrom, H., Fontes, Ed. *Electrochemistry Communications* 2014, 40, 71.
[11] Moya, A. A. *J. Phys. Chem. C* 2014, 118, 2539.
[12] Moya, A. A. *Electrochim. Acta* 2010, 55, 2087.
[13] Moya, A. A. *Electrochim. Acta* 2011, 56, 3015.
[14] Moya, A. A. *Electrochim. Acta* 2012, 62, 296.
[15] Moya, A. A. *Electrochim. Acta* 2013, 90, 1.
[16] Horno, J., Moya, A. A., González-Fernández, C. F. *J. Electroanal. Chem.* 1996, 402, 73.
[17] Franceschetti, D. R., Macdonald, J. R., Buck, R. P. *J. Electrochem. Soc.* 1991, 138, 1368.
[18] Buck, R. P. *Electrochim. Acta* 1993, 38, 1837.
[19] Buck, R. P., Mundt, C. *J. Chem. Soc. Faraday Trans.* 1996, 92, 3947.
[20] Danilov, D., Notten, P. H. L. *Electrochim. Acta* 2008, 53, 5569.
[21] Danilov, D., Notten, P. H. L. *J. Power Sources* 2009, 189, 303.
[22] Moya, A. A., Castilla, J., Horno, J. *J. Phys. Chem.*, 1995, 99, 1292.
[23] Moya, A. A., Hayas, A., Horno, J. *J. Electroanal. Chem.* 1996, 413, 1.
[24] Moya, A. A., Hayas, A., Horno, J. *J. Electroanal. Chem.* 1996, 413, 9.
[25] Barsoukov, E., Macdonald, J. R. *Impedance Spectroscopy: Theory, Experiment and Applications*; Wiley: New York, 2005.
[26] Brumleve, T. R., Buck, R. P. *J. Electroanal. Chem.* 1981, 126, 73.
[27] Diard, J.-P., Le Gorrec, B., Montella, C. *J. Electroanal. Chem.* 1999, 471, 126.

In: Computational Chemistry
Editor: Daria Bove
ISBN: 978-1-63321-354-8

Chapter 4

MOLECULAR SIMULATION OF ELECTRON BEAM NANOFABRICATION

Masaaki Yasuda[1,*] ***and Kazuhiro Tada***[2]
[1]Graduate School of Engineering, Osaka Prefecture University, Sakai, Japan
[2]National Institute of Technology, Toyama, Japan

ABSTRACT

Electron beam irradiation is expected to become a technique for tailoring the structure of materials to obtain desirable properties. Molecular dynamics (MD) simulations are a powerful tool to understand the structural changes of materials at the atomic scale. In this chapter, MD simulations of electron beam nanofabrication are introduced. Electron beam modification of carbon nanomaterials, the pattern formation process in electron beam lithography, and electron irradiation damage in silicon oxides are investigated with MD simulations.

1. INTRODUCTION

The effects of electron irradiation on materials are important in various fields, such as nuclear engineering, space technology, micro-beam analyses,

* E-mail:yasuda@pe.osakafu-u.ac.jp (M. Yasuda).

electron microscopy, and nanofabrication. In the case of electron microscopy, electron irradiation damage of samples becomes a serious problem for accurate structure observation and decreases the accuracy of size measurements. In nanoscience and nanotechnology, transmission electron microscope (TEM) observations are indispensable for the characterization of nanomaterials. However, the electrical, mechanical, and other properties obtained from TEM observations are affected by electron irradiation. Furthermore, in-situ observations of chemical reactions and crystal growth at the atomic scale become possible using an environmental TEM. Even with such observations, there is a possibility that electron irradiation affects the observed phenomena.

Despite the negative aspects of irradiation effects, energetic beam irradiation of materials is expected to become a technique for tailoring the structure to obtain desirable properties [1, 2]. Taking carbon nanomaterials as an example, several experiments have structurally modified carbon nanomaterials.

A bundle of single-walled carbon nanotubes (SWNTs) was cut with a focused electron beam [3]. Electron-irradiated multi-walled carbon nanotubes were shown to bend because of the removal of carbon atoms from one side of the tube [4]. Kink structures were obtained in SWNTs through the sequential removal of atoms by electron irradiation [5].

Furthermore, nanotube welding by electron irradiation has been demonstrated, and various “X”, “Y”, and “T” junctions of SWNTs were created [6]. Coalescence of fullerenes inside carbon nanotubes by electron irradiation has also been reported [7]. Graphene nanostructures have been sculpted by electron beam irradiation [8]. A carbon atomic chain has been obtained by removing carbon atoms row by row from graphene using electron irradiation [9]. However, such structural modifications of carbon nanomaterials with electron beams are not well-established at present. Although several theoretical studies have investigated irradiation-induced structural changes in carbon nanomaterials [10-14], further investigation is required.

In the case of polymer materials, electron beam lithography (EBL) is one of the typical applications of electron irradiation. In EBL, exposure to an electron beam causes chemical changes in the electron-sensitive polymer materials (electron resist). The solubility of the electron-exposed resist in a specific solvent changes. Using this property in the development process, a nanometer-scale pattern can be created on the surface of the material.

Pattern formation by EBL is greatly affected by electron scattering in the materials. Therefore, numerical simulations, such as Monte Carlo simulations of the electron scattering, are indispensable for predicting pattern profiles and correcting proximity effects in EBL [15-19]. However, the required feature sizes of patterns are becoming smaller in the development of nanodevices. With the decrease in pattern size, the behavior of the polymer molecule in the resist becomes crucial to understand the pattern formation process. Because conventional simulations for EBL do not consider the resist's molecular behavior, the accuracy of the simulations decreases.

In this chapter, we report molecular dynamics (MD) simulations including electron irradiation effects based on a Monte Carlo method to study the nanofabrication process of materials with electron beams. MD simulations are a powerful tool to reveal the atomic scale behavior of materials. We introduce the interaction between an incident electron and the material as a stochastic process in the MD simulations, and study the structural changes of the materials under electron irradiation.

First, we describe the simulation model and explain how we introduce the electron irradiation effects into the MD simulations.

Second, we apply the simulation method to carbon nanomaterials. The electron irradiation damages under TEM observation, nanofabrication with electron beams, and the variation of the mechanical properties by electron irradiation are demonstrated.

Third, we apply the method to polymer materials. We investigate pattern formation in EBL using MD simulations and irradiation damage under scanning electron microscope (SEM) observation. We include application to irradiation damage under electron microscope observation because electron microscopy is an important technique to investigate the nanostructure in nanofabrication.

Finally, we apply the method to the irradiation effects in silicon oxide.

2. Methodology

In the present simulations, we include electron irradiation effects in classical MD simulations. The electron collision was classified as either elastic or inelastic one. In the present study, we included an elastic collision for the carbon nanomaterials and silicon oxide, and an inelastic collision for the polymer materials.

Figure 1 shows the elastic collision model of the incident electrons. To determine the scattering angle of incident electron ω, the following screened Rutherford cross section was used [20]:

$$\frac{d\sigma}{d\Omega}=\frac{e^4 Z(Z+1)}{m_e^2 v^4 (1-\cos\omega+2\beta)^2}, \tag{1}$$

where m_e is the electron mass, v is the electron velocity, e is the electron charge, Z is the atomic number of the target atom, and β is the screening parameter. When the target atom is a heavy element, using the Mott cross section increases the precision of the simulation [21]:

$$\frac{d\sigma}{d\Omega}=|f(\omega)|^2+|g(\omega)|^2, \tag{2}$$

where $f(\omega)$ and $g(\omega)$ are the scattering amplitudes.

After determining the scattering angle of the incident electron, the energy transferred from the electron to the target atom E_t and the scattering angle of the target atom from the incident axis θ were obtained based on the relativistic binary collision model as follows:

$$E_t=\left\{1-\left(\frac{m_e\cos\omega+\sqrt{m_t^2-m_e^2\sin^2\omega}}{m_e+m_t}\right)^2\right\}\frac{\left(E_0+2m_e c^2\right)}{2m_e c^2}E_0, \tag{3}$$

$$\cos\theta=\sqrt{\frac{(m_e+m_t)^2}{4m_e m_t}\cdot\frac{E_t}{E_0}}, \tag{4}$$

where m_t is the target atom mass, E_0 is the incident electron energy, and c is the speed of light. The scattering angle of the target atom around axis ϕ is uniformly distributed.

The collision atom was randomly selected in the irradiated area of the target materials. The motions of the target atoms were calculated with MD simulations introducing the above-mentioned electron irradiation effect every 2000 MD steps. The time step used in the MD simulations was 0.5 fs.

The multiple scattering of the electrons in the material was not considered in the present simulations with the assumption that the size of the material was sufficiently small for the incident electron energy. In actual electron irradiation, electron excitations occur and cause structural changes in the target materials. However, electron excitations were not included in the above-mentioned model. The current density of the electron beam in our simulations was estimated to be about 10 orders of magnitude larger than that in TEM experiment. Because the cross section of the collision that causes serious structural change is quite small, the target material was sufficiently relaxed between the irradiation steps [22].

Figure 2 shows the introduction of the inelastic electron collision effects. Because the actual inelastic collision process of the incident electrons is complicated, we included the irradiation effects through the loss of electron energy in the target materials. The electron trajectories in the target material were calculated by Monte Carlo simulation of the electron scattering. Then, the target material was divided into small regions. The absorbed energy in each region was obtained from the loss of electron energy along the electron trajectories with the Monte Carlo simulation. The electron irradiation effect was included according to the absorbed electron energy in each region.

In the Monte Carlo simulation [23], we used the screened Rutherford cross section [20] and the Bethe equation [24] to calculate the elastic scattering and the electron energy loss. The generation of the secondary electrons in the target materials was not included in the present simulations.

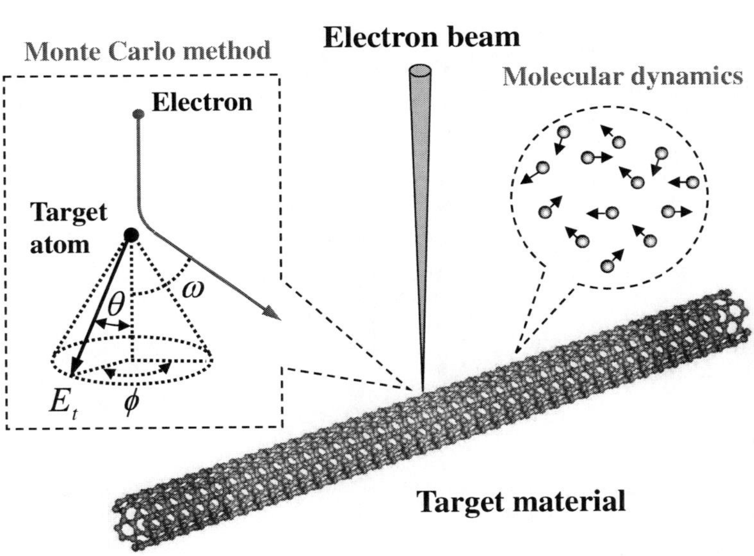

Figure 1. Elastic collision model of the incident electrons. The momentum transferred from the electron to the target atom was included based on the binary collision model using an elastic collision cross section.

Taking the polymer material as an example, the effect of electron irradiation was included by chain scission of the polymer molecule in the MD simulation. The rate of chain scission in the material was set to be proportional to the absorbed energy in each region preliminarily calculated by Monte Carlo simulation of the electron scattering. The breaking positions in the polymer chain were randomly selected. Side-chain scission of the polymer molecule and the emission of the decomposition gases can also be included as an irradiation effect. Chain scission and the structural relaxation processes were alternately repeated in the exposure simulation.

The motion of each atom in the target material was calculated by MD simulation using empirical potentials. For the carbon nanomaterials, the short- and long-range interactions were described by Tersoff-Brenner [25-27] and Lennard-Jones potentials, respectively. Both potentials are smoothly connected with cubic splines [28].

The force field reported by Okada et al. [29] was used for the polymer materials. It consists of bond stretching, bending, torsion potentials, and nonbonding interaction including Lennard-Jones and Coulomb potentials.

For the silicon oxide materials, the Born-Mayer-Huggins potential including a three-body term was used to describe the interactions between atoms [30].

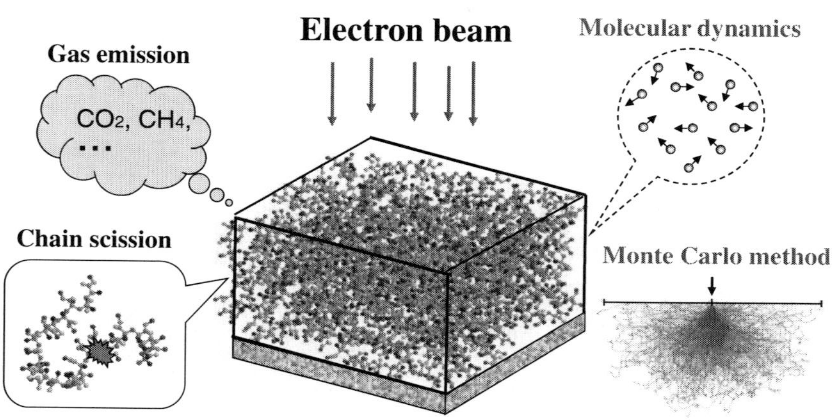

Figure 2. Introduction of the inelastic electron collision effects. The effect of electron irradiation was included by chain scission of the polymer molecule or decomposition gas emission in the MD simulation. The effect was set to be proportional to the absorbed energy distribution preliminarily calculated by Monte Carlo simulation of the electron scattering.

3. SIMULATION RESULTS

The structural changes of the materials under electron irradiation are largely affected by the irradiation conditions. To tailor structures by electron irradiation, understanding the relationship between the irradiation conditions and the structural changes is important. Here, we performed MD simulation of the electron beam nanofabrication process for various conditions and illustrate typical examples of the structural changes of carbon nanomaterials, polymer materials and silicon oxide by electron irradiation.

3.1. Carbon Nanomaterials

3.1.1. Electron Irradiation Damages

Electron irradiation causes several types of structural changes, such as knock-on, pentagon-heptagon (5-7), and Stone-Wales defect. Figure 3 shows the variation of the potential energy of graphene before and after 5-7 defect formation by electron irradiation. The electron energy was 200 keV. The sharp peak in the potential energy shows the electron collision point. When electron collision provides more energy than the activation energy, a 5-7 defect is formed in the honeycomb network of graphene. The potential energy of the structure increased by about 5 eV after the formation of a 5-7 defect.

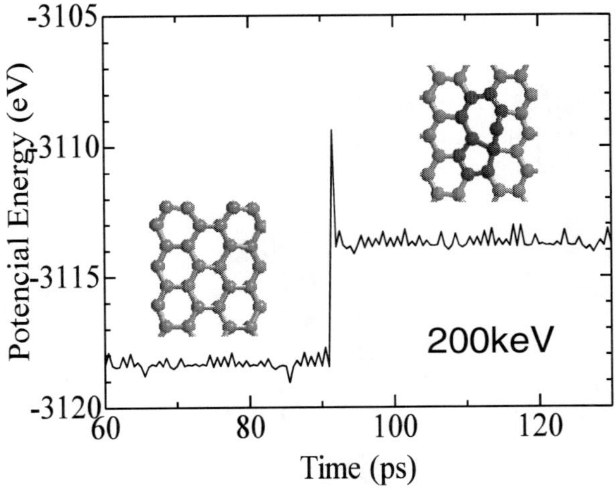

Figure 3. Variation of the potential energy of graphene before and after formation of a 5-7 defect by electron collision at 200 keV.

Typical examples of the electron irradiation defects observed in graphene are shown in Figure 4. When the electron-irradiated atom receives sufficient energy to break all the chemical bonds with surrounding atoms, it is ejected from the structure as a knock-on atom (Figure 4(a)), leaving a vacancy in the structure. When the electron-irradiated atom changes its bonding partner to a neighboring atom, a 5-7 defect is formed in the honeycomb network (Figure 4(b)). When the atom is punched out of the structure by the electron collision, it frequently adheres to surrounding atoms as an adatom. As a result, an adatom-vacancy pair is formed in the structure (Figure 4(c)). Stone-Wales defects also appear as an electron irradiation defect, and consist of double pentagon-heptagon pairs that result from 90° bond rotation in the honeycomb network (Figure 4(d)).

The energy dependence of the electron irradiation damage of SWNTs was investigated with MD simulations. Figure 5 shows the structures of (5,5) SWNTs obtained by MD simulations after electron irradiation at several electron energies. The tube length was 5 nm and both ends of the nanotubes were fixed.

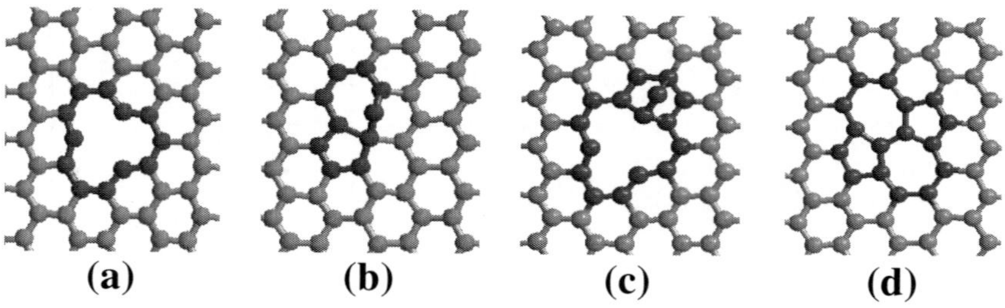

Figure 4. Typical examples of the electron irradiation defects observed in graphene. (a) Knock-on defect, (b) pentagon-heptagon pair, (c) adatom-vacancy pair, and (d) Stone-Wales defect.

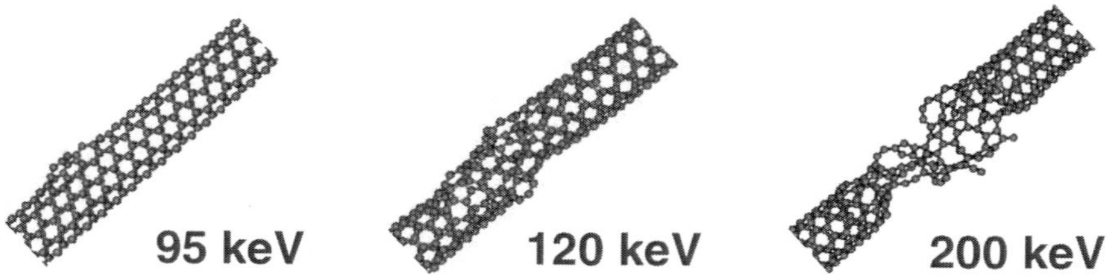

Figure 5. Structures of electron-irradiated (5,5) SWNTs at different electron energies. With increasing electron energy, the irradiation damage became more serious.

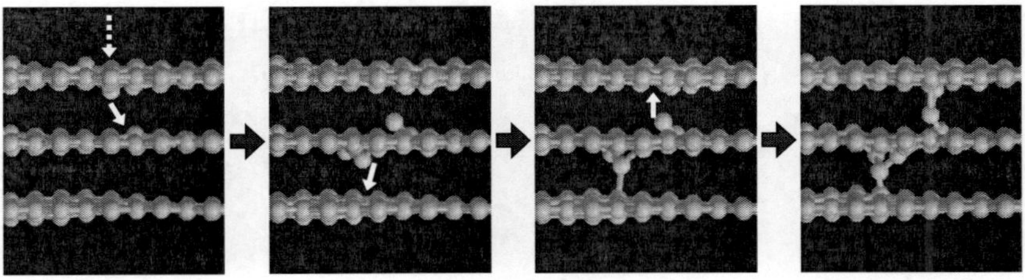

Figure 6. Collision cascade process in electron-irradiated trilayergraphene. Only the top layer was irradiated by an electron. Two crosslinks are formed between the layers by a single electron collision.

The center 2-nm-long region was irradiated by the electrons. Irradiation damage was observed above electron energy of 95 keV in the present simulations. With increasing electron energy, the irradiation damage of the SWNTs became more serious.

Figure 6 shows the collision cascade process observed in trilayergraphene. A 200 keV electron collides with a carbon atom in the top layer. The collision atom is punched out from the top layer and collides with another atom in the middle layer. This collision punches out the atom from the middle layer and forms a crosslink between the middle and bottom layers. The punched-out atom from the top layer is backscattered by the collision with the middle layer and another crosslink is formed between the top and the middle layers. As a result, two crosslinks are formed between the layers by a single electron collision.

3.1.2. Electron Beam Fabrication

By controlling the structural change, electron irradiation becomes a fabrication technique of carbon nanomaterials. First, cutting of the SWNT by a focused electron beam was simulated [31]. The cutting process for the (5,5) SWNT by 200 keV electron irradiation is shown in Figure 7. The electron beam spot is shown by the circle in the figure and the diameter of the spot is 1 nm. Both ends of the SWNT were fixed. The temperature of both ends was kept constant at 1000 K. At the first stage of cutting, carbon atoms are randomly ejected by the electron collision. Then, the SWNT is gradually bisected. Because high temperature promotes the recombination of dangling bonds, the cut edges of the SWNTs begin to close. Finally, the SWNT is perfectly cut, and both cut edges are self-capped.

The bending deformation of SWNT by electron irradiation was also simulated [31]. Figure 8 shows bending deformations of (8,8) SWNTs by 200 keV electron irradiation.

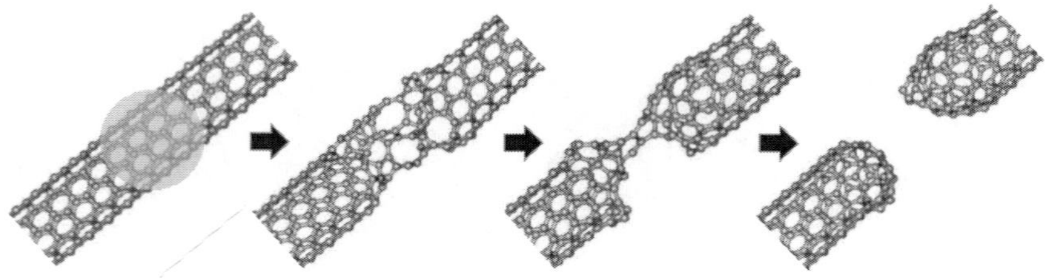

Figure 7. Cutting process of a (5,5) SWNT by 200 keV focused electron beam irradiation at 1000 K. The electron beam spot is shown by the circle.

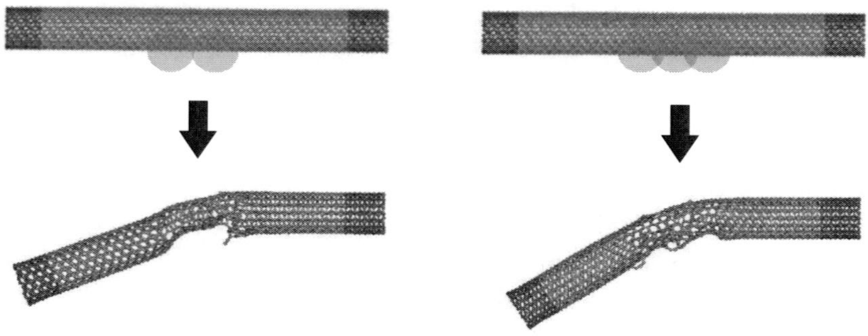

Figure 8. Bending deformations of (8, 8) SWNTs by 200 keV electron irradiation. The parts highlighted with the circles were irradiated with electrons.

The tube length was 10 nm and one end of the SWNT was fixed. The tube temperature was kept constant at 1000 K. The parts of the SWNTs highlighted by the circles were irradiated with electrons. Two results with different irradiated areas are shown. Some carbon atoms were ejected after collision with the incident electrons. The structural transition to an amorphous-like structure in the irradiated area can be observed in the figures. The irradiated parts of the SWNTs consist of many pentagons and heptagons. Because the number of carbon atoms decreased after ejection, the irradiated part of the SWNT shrank, causing contraction stress in the SWNT and resulting in bending of the SWNT. The bending angle increased with an increase in the width of the irradiated area.

Welding of the SWNTs by electron irradiation has also been reported [6]. Figure 9 shows the simulation of the welding process of two (10, 10) SWNTs by 150 keV electron irradiation. The temperature of both SWNTs was kept constant at 1000 K. Both SWNTs have cap structures at their ends. In the simulation, two SWNTs were welded and become one SWNT after electron irradiation. In the welded region, a tubular structure is observed. However, the tube wall is no longer composed of the hexagonal honeycomb lattice. The structure in this region is amorphous-like and has many defects caused by electron irradiation.

Constriction is observed in the welded region in the SWNT. Experimentally, seamless connection is observed when the same types of SWNTs are welded [32]. Although the process conditions are not optimized in the simulation, seamless welding of nanotubes was difficult to attain. Further investigation into the details is required for the welding process.

Graphene nanostructures fabricated by electron beam irradiation have also been reported [8]. The most typical irradiation defect observed in layered graphene is crosslinking between the layers. By applying this crosslink formation, three dimensional structures can be created from bilayer graphene [33]. Figure 10 shows the formation of a nanotube junction from bi-layer graphene by punching four adjacent holes with electrons. The four areas as shown by squares were irradiated with 500 keV electrons.

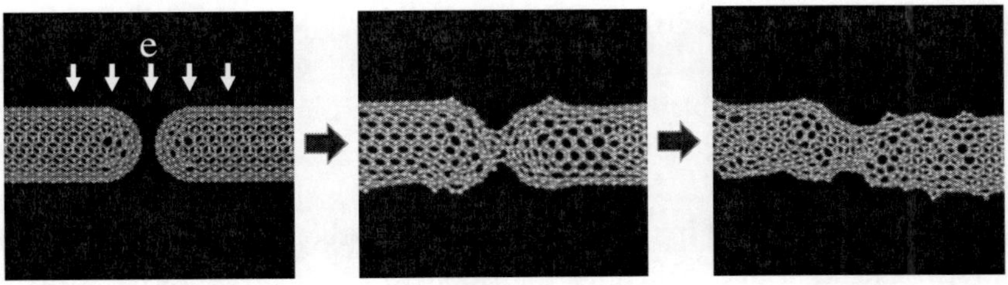

Figure 9. Welding process of two (10, 10) SWNTs by 150 keV electron irradiation. The structure of the welded region is amorphous-like and constriction of the region is observed.

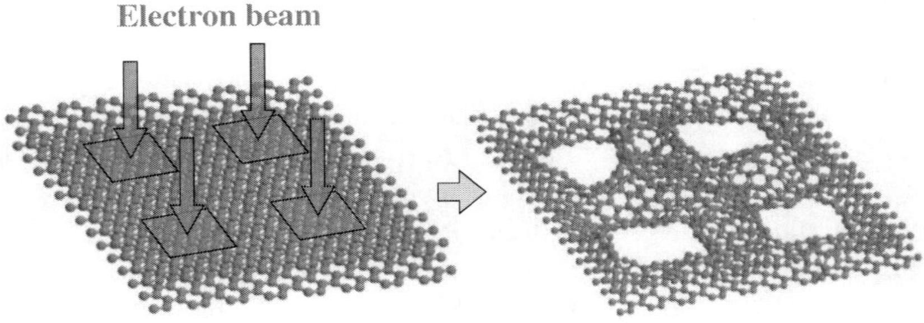

Figure 10. Formation of a nanotube junction from bi-layer graphene by 500 keV electron irradiation. The structure was annealed at 1500 K after irradiation. A nanotube junction is observed between the four punched holes.

The structure was annealed at 1500K after irradiation. Because the rims of the four holes are joined through atomic bonds, tubular structures are formed between the holes. As a result, a nanotube junction is observed in the bi-layer graphene.

3.1.3. Nanomechanics

Electron irradiation is a technique for modifying the mechanical properties of nanomaterials. In this section, the mechanical properties, such as the tensile and torsional properties, of electron-irradiated SWNTs are investigated with MD simulations [34]. Figure 11 shows schematic views of the structural change of a (6, 6) SWNT irradiated with 200 keV electrons at 300 and 2500 K. The SWNT was 5 nm long. Figures 11(a) and (e) show the initial structure. At 300 K, after 5 ns irradiation, the tube structure begins to break up, some carbon atoms are ejected from the SWNT, and nonhexagonal rings, such as pentagons and heptagons, are formed in the tube wall (Figure 11(b)). After 15 ns irradiation, the number of ejected atoms increases and the tubular structures are collapsing (Figure 11(c)). After 40 ns irradiation, the SWNT cracks and the structure is no longer cylindrical (Figure 11(d)). Conversely, the SWNT maintains a cylindrical shape at 2500 K (Figure 11(f)-(h)). However, the tube wall is not composed of a hexagonal honeycomb network. Because of the high temperature, surface reconstruction of the tube wall is promoted through dangling bond saturation in the SWNT after the carbon atoms are removed by electron irradiation and structural transition to an amorphous-like structure occurs.

Figure 12 shows the stress-stain characteristics for tension tests of the (6, 6) SWNT after 200 keV electron irradiation at (a) 300 and (b) 2500 K. The SWNT was 5 nm long. After 5 ns irradiation at 300 K, the yield stress and strain decrease compared with the pristine SWNT. The degradation corresponds to the electron irradiation defects shown in Figure 11(b). The Young's modulus decreases and the strain increases with increasing irradiation time. This is because the SWNT grows complacent as the hexagonal honeycomb lattice changes to nonhexagonal rings, such as heptagons, and several carbon chains, as shown in Figs. 11(c) and (d). However, the variation of the stress and strain after irradiation at 2500 K is small compared with the SWNT irradiated at 300 K. Because the SWNT maintains the shape of the cylinder, as shown in Figures 11(f)-(h), the nanotube irradiated at high temperature is robust.

Figure 13 shows torque as a function of torsional angle for (6,6) SWNTs after 200 keV electron irradiation at (a) 300 and (b) 2500 K. For the tube irradiated at 300 K, the yield point disappears and the torque decreases as the irradiation time increases. Conversely, for the tube irradiated at 2500 K, the torque stays about the same at the different irradiation times. Similar to the tensile property, the torsional properties of the SWNT irradiated at the higher temperature are stronger than those of the SWNT irradiated at the lower temperature.

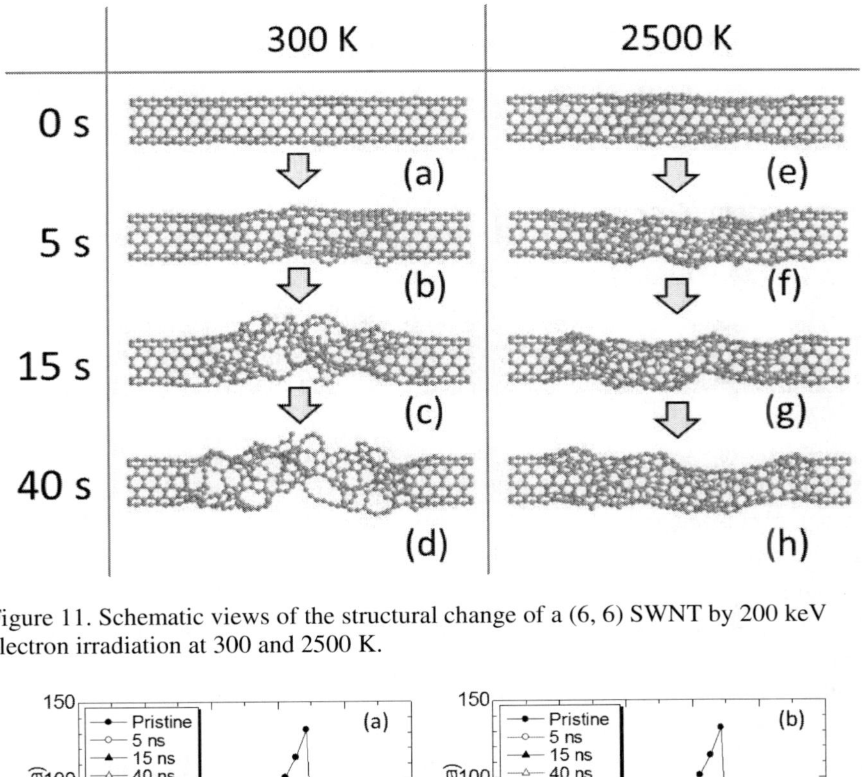

Figure 11. Schematic views of the structural change of a (6, 6) SWNT by 200 keV electron irradiation at 300 and 2500 K.

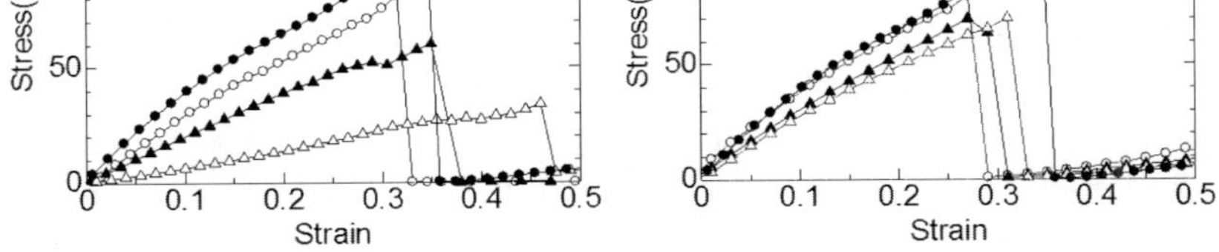

Figure 12. Stress-stain characteristics for tension tests of the (6, 6) SWNT after 200 keV electron irradiation at (a) 300 and (b) 2500 K.

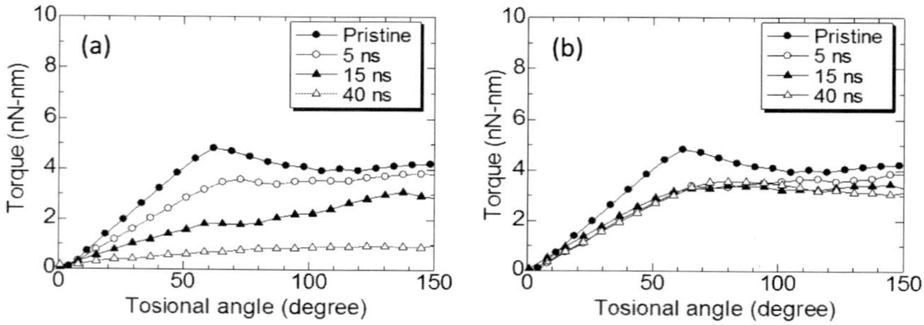

Figure 13. Relationships between torque and torsional angle of (6,6) SWNTs irradiated at (a) 300 and (b) 2500 K.

3.2. Polymer Materials

3.2.1. Electron Beam Lithography

Here, we applied our simulation to the pattern formation process in EBL. The pattern formation in EBL consists of two processes: electron exposure and resist development [35, 36]. The effect of electron exposure was included in the MD simulation by main-chain scission of the PMMA polymer. The cutting positions in the polymer chain were randomly selected. Chain scission and the structural relaxation processes were alternately repeated in the MD simulation. The sample used in the simulation was poly(methyl methacrylate) (PMMA) resist on a Si substrate. The film thickness of the PMMA resist was 4 nm. The molecular weight of the initial PMMA polymer was 5000.

Figure 14 shows snapshots of the PMMA resist molecules under electron exposure obtained from the MD simulation. One PMMA molecule is highlighted as a marker molecule, which was segmented as the electron exposure progressed. Each segment of the molecule slightly moved from its original position. As the electron exposure proceeded, the original shape of the marker molecule was hardly recognizable from the constituent segments.

When the patterned exposure in EBL was simulated, the chain scission rate was set to be proportional to the absorbed energy distribution in the resist calculated by Monte Carlo simulation of the electron scattering. The contribution of the backscattered electrons from the substrate was included in the absorbed energy distribution. Secondary electron generation was not considered in the present Monte Carlo simulation.

The development process was modeled by removing small segments of polymer molecule from the resist structure in ascending order (from small to large molecules). The resist developed from the surface to the bottom layer, as shown in Figure 15. First, resist molecules were removed from the surface layer in ascending order. Monomers were first removed from the resist. Then, dimers, trimmers, and tetramers were sequentially removed. Molecule removal and the structural relaxation processes were alternately repeated in the MD simulation. After development of the surface layer finished, the second layer was developed in the same manner. The same process was repeated until the bottom layer was developed. The temperature of the sample was kept constant at 300 K in both the electron exposure and the development process.

Figure 16 shows the development process of the 2-nm-wide line pattern exposed at 100 kV obtained from the simulation. Molecular segments whose polymerization degree was less than or equal to 4 were removed from the structure. The substrate was gradually exposed as development proceeded. Molecular-scale roughness was observed on the sidewalls of the pattern. The presence and absence of the molecular chain are observed as a typical structure of the atomic-scale line edge roughness.

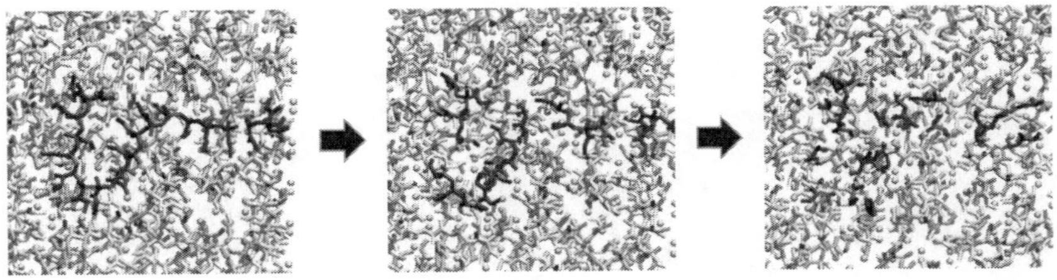

Figure 14. Snapshots of the PMMA resist molecules under electron exposure. One PMMA molecule is highlighted as a marker molecule.

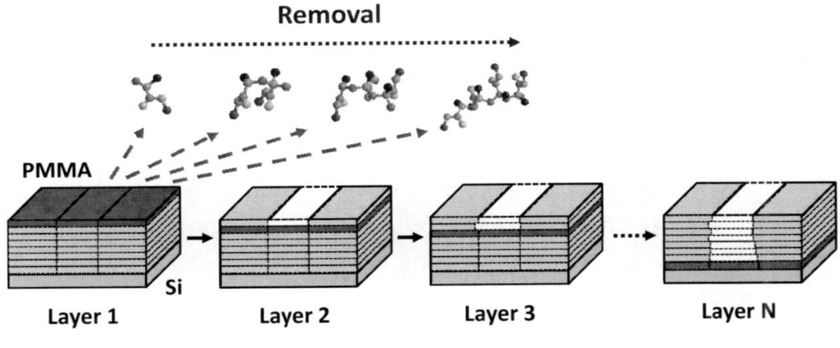

Figure 15. Simulation model of the development process in EBL. Small segments of the polymer molecule were removed from the resist structure.

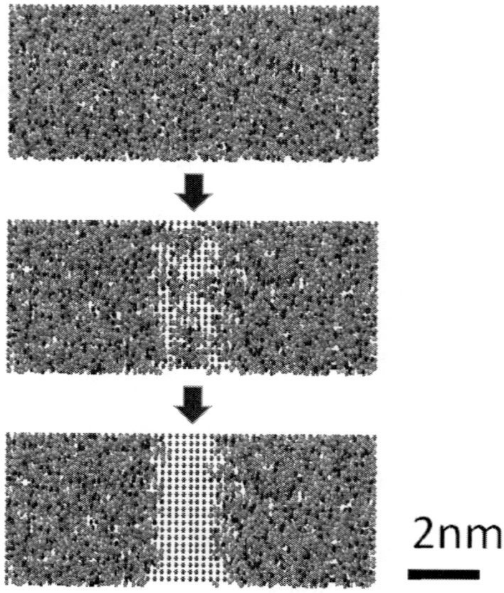

Figure 16. Example of the 2-nm-wide line pattern exposed at 100 kV. Molecular-scale roughness is observed on the sidewalls of the pattern.

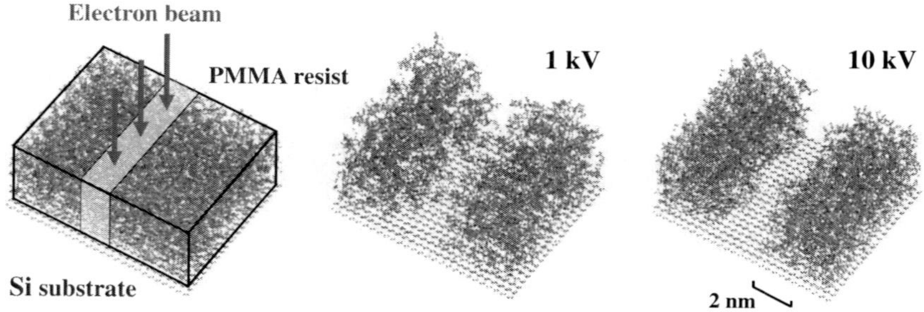

Figure 17. Resist pattern structures exposed at accelerating voltages of 1 and 10 kV. The edge shapes of the patterns exposed at 10 kV are smoother than those at 1 kV.

The degree of the roughness largely depends on the accelerating voltage. Figure 17 shows the resist pattern structures exposed at accelerating voltages of 1 and 10 kV. Molecular segments whose polymerization degree was less than or equal to 4 were removed from the structure. The edge shapes of the patterns exposed at 10 kV are smoother than those at 1 kV.

3.2.2. Pattern Shrinkage in Scanning Electron Microscopy

Electron irradiation damage is also a serious problem in SEM observation of organic polymer materials. The resist patterns frequently shrink with electron irradiation [37, 38]. This shrinkage decreases the accuracy of the pattern size measurement using critical-dimension SEM. Here, we performed a MD simulation to investigate resist pattern shrinkage under electron irradiation.

PMMA resist on a Si substrate was selected as the sample material. The molecular weight of the PMMA molecule was 50,000. The effect of electron irradiation was included by side-chain scissions of the PMMA polymer in the MD simulation. In the actual chemical reaction, main-chain scission also occurs under electron irradiation. However, main-chain scission was not included in the present simulation. The breaking positions in the polymer chain were randomly selected. The decomposition gases, such as CO_2 and CH_4, resulting from side-chain scission were removed from the structure. The chain scission and structural relaxation processes were alternately repeated in the simulations.

Figure 18 shows examples of the structural changes of the PMMA resist with electron irradiation. The initial size of the PMMA resist was $6 \times 6 \times 6$ nm^3. 50% and 100% of the side chains were removed. Shrinkage of the resist caused by side-chain scission and decomposition gas emission was observed.

Figure 18. Structural change of the PMMA resists by electron irradiation.

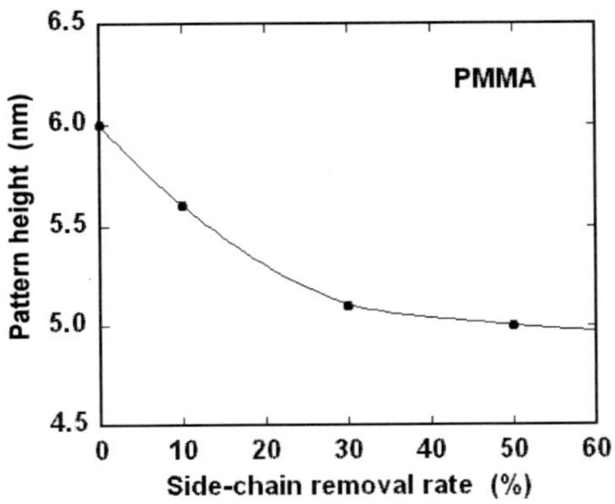

Figure 19. Pattern height of the PMMA resist as a function of side-chain removal rate.

Figure 19 shows the pattern height of the PMMA resist as a function of side-chain removal rate obtained from the MD simulation. The pattern height decreases with increasing removal rate. The MD simulation only considers a small size of resist because of computational time. To analyze the shrinkage of larger patterns, the pattern was divided into small regions and the shrinkage of each region was estimated from the absorbed energy in each region calculated by Monte Carlo simulation of the electron scattering.

3.3. Silicon Oxide

Finally, to apply the present simulation to an inorganic material, we performed a MD simulation of the electron irradiation process for SiO_2. The target SiO_2 volume was $4 \times 5 \times 3$ nm^3. The electron irradiation effect was included by the binary collision model using an elastic collision cross section. Inelastic collision effects were not included.

Figure 20 shows the structural change of crystalline SiO_2 (c-SiO_2) under 1.0 MeV electron irradiation. The electron beam spot is shown by the circle. As the irradiation time increased, the structure in the irradiation region changed from crystalline to amorphous. The amorphous region increases with increasing irradiation time. Because the Si and O atoms are driven into the slab, a slight dimple is observed at the surface.

Figure 21 shows the variation of the Si composition ratio in the surface region of c-SiO_2 (Figure 21(a)) and amorphous SiO_2 (a-SiO_2) (Figure 21(b)) with time for various electron energies. As the electron energy increased, the Si composition ratio in the surface increased for both c-SiO_2 and a-SiO_2. This is because oxygen atoms are more easily driven into the slab than silicon atoms because of their lighter atomic mass. This tendency becomes more significant with increasing irradiation time.

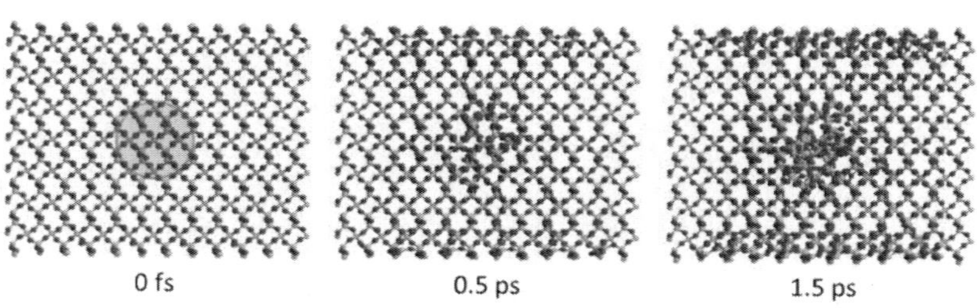

Figure 20. Structural changes of c-SiO_2 under 1.0 MeV electron irradiation. The electron beam spot is shown by the circle.

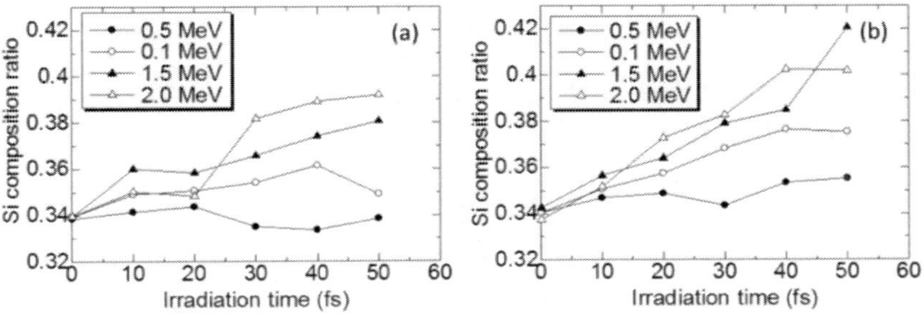

Figure 21. Variation of the Si composition ratio in the surface region with time for various electron energies: (a) c-SiO_2 and (b) a-SiO_2.

CONCLUSION

MD simulations were performed to study the structural changes of materials by electron irradiation. Electron irradiation effects were included by the Monte Carlo method. The electron elastic collision was modeled by the binary collision model using an elastic collision cross section. The inelastic collision effect was included through the absorbed energy distribution in the materials obtained by Monte Carlo simulation of the electron scattering. Structural changes were included in proportion to the absorbed energy distribution.

In the analysis of the structural changes of carbon nanomaterials, several types of defect structures were revealed. Cutting, bending, welding, and sculpting of the materials by electron irradiation was demonstrated. The change of the mechanical properties by electron irradiation was also shown. For application of the simulations to polymer materials, 2-nm-wide pattern formation in EBL was simulated. The structure of the atomic-scale line edge roughness was shown. The resist pattern shrinkage in SEM observation is also investigated. Finally, electron irradiation damage of silicon oxide was investigated. The electron energy dependence of the structural changes was determined.

Although further improvement is required to increase the precision of the simulation method, the present simulations provide useful information to understand the electron irradiation effects in materials.

Acknowledgment

This work was supported by the Japan Society for the Promotion of Science (JSPS) KAKENHI (Grant Numbers 22360145, 23656245, and 25249052).

References

[1] A. V. Krasheninnikov and F. Banhart, *Nature Mater*. 6, 723 (2007).
[2] A. V. Krasheninnikov and K. Nordlund, *J. Appl. Phys.* 107, 071301 (2010).
[3] F. Banhart, J. Li, and M. Terrones, *Small 1*, 953 (2005).
[4] J. Li and F. Banhart, *Nano Lett*. 4, 1143 (2004).
[5] A. Zobelli, A. Gloter, C. P. Ewels, and C. Colliex, *Phys. Rev. B* 77, 045410 (2008).
[6] M. Terrones, F. Banhart, N. Grobert, J.-C. Charlier, H. Terrones, and P. M. Ajayan, *Phys. Rev. Lett*. 89, 075505 (2002).
[7] D. E. Luzzi and B. W. Smith, *Carbon* 38, 1751 (2000).
[8] B. Song, G. F. Schneider, Q. Xu, G. Pandraud, C. Dekker, and H. Zandbergen, *Nano Lett.* 11, 2247 (2011).
[9] C. Jin, H. Lan, L. Peng, K. Suenaga, and S. Iijima, *Phys. Rev. Lett*. 102, 205501 (2009).
[10] P. M. Ajayan, V. Ravikumar, and J.-C. Charlier, *Phys. Rev. Lett.* 81, 1437 (1998).
[11] S. K. Pregler and S. B. Sinnott, *Phys. Rev. B* 73, 224106 (2006).
[12] A. V. Krasheninnikov, F. Banhart, J. X. Li, A. S. Foster, and R. M. Nieminen, *Phys. Rev. B* 72, 125428 (2005).
[13] O. V. Yazyev, I. Tavernelli, U. Rothlisberger, and L. Helm, *Phys. Rev. B* 75, 115418 (2007).
[14] I. Jang, S. B. Sinnott, D. Danailov, and P. Keblinski, *Nano Lett.* 4, 109 (2004).
[15] K. Murata, D. F. Kyser, and C. H. Ting, *J. Appl. Phys*. 52, 4396 (1981).
[16] G. Owen and P. Rissman, *J. Appl. Phys.* 54, 3573 (1983).
[17] M. Kotera, T. Yamada, and Y. Ishida, *Jpn. J. Appl. Phys.* 41, 4150 (2002).
[18] N. Tsikrikas, D. Drygiannakis, G. P. Patsis, I. Raptis, S. Stavroulakis, and E. Voyiatzis, *Jpn. J. Appl. Phys.* 46, 6191 (2007).

[19] M. Stepanova, T. Fito, Z. Szabó, K. Alti, A. P. Adeyenuwo, K. Koshelev, M. Aktary, and S. K. Dew, *J. Vac. Sci. Technol. B* 28, C6C48 (2010).

[20] G. Wentzel, *Z. Phys.* 40, 590 (1927).

[21] N. F. Mott, *Proc. R. Soc. Lond. Ser. A* 124, 425 (1929).

[22] M. Yasuda, Y. Kimoto, K. Tada, H. Mori, S. Akita, Y. Nakayama, and Y. Hirai, *Phys. Rev. B* 75, 205406 (2007).

[23] K. Murata, E. Nomura, K. Nagami, T. Kato, and H. Nakata, *Jpn. J. Appl. Phys.* 17, 1851 (1978).

[24] H. A. Bethe, *Handb. Phys.* 24, 273 (1933).

[25] J. Tersoff, *Phys. Rev. B* 37, 6991 (1988).

[26] J. Tersoff, *Phys. Rev. B* 39, 5566 (1989).

[27] D. W. Brenner, *Phys. Rev. B* 42, 9458 (1990).

[28] D. W. Brenner, D. H. Robertson, M. L. Elert, and C. T. White, *Phys. Rev. Lett.* 70, 2174 (1993).

[29] O. Okada, K. Oka, S. Kuwajima, S. Toyoda, and K. Tanabe, *Comput. Theor. Polym. Sci.* 10, 371 (2000).

[30] J. M. Delaye, V. Louis-Achille, and D. Ghaleb, *J. Non-Cryst. Solids,* 210, 232 (1997).

[31] M. Yasuda, R. Mimura, H. Kawata, and Y. Hirai, *J. Appl. Phys.* 109, 054304 (2011).

[32] C. Jin, K. Suenaga, and S. Iijima, *Nature Nanotechnol.* 3, 17 (2008).

[33] Y. Asayama, M. Yasuda, K. Tada, H. Kawata, and Y. Hirai, *J. Vac. Sci. Technol. B* 30, 06FJ02 (2012).

[34] K. Tada, M. Yasuda, T. Mitsueda, R. Honda, H. Kawata, and Y. Hirai, *Microelectron. Eng.* 107, 50 (2013).

[35] M. Yasuda, H. Sakai, R. Takai, H. Kawata, and Y. Hirai, *Microelectron. Eng.* 112, 287 (2013).

[36] K. Michishita, M. Yasuda, H. Kawata, and Y. Hirai, *Jpn. J. Appl. Phys.* 53, 06JB02 (2014).

[37] T. Azuma, K. Chiba, H. Abe, H. Motoki, and N. Sasaki, *J. Vac. Sci. Technol. B* 22, 226 (2004).

[38] T. Ohashi, T. Sekiguchi, A. Yamaguchi, J. Tanaka, and H. Kawada, *Jpn. J. Appl. Phys.* 52, 06GB03 (2013).

In: Computational Chemistry
Editor: Daria Bove
ISBN: 978-1-63321-354-8

Chapter 5

INTERPRETATION OF CHIRAL ORDERING OF HYBRID SYSTEM OF SEVERAL AZO DYES AND CHIRAL SCHIFF BASE CO(II) COMPLEX INDUCED BY CIRCULARLY POLARIZED LIGHT

Nobumitsu Sunaga, Shohei Furuya, Maiko Ito, Chigusa Kominato and Takashiro Akitsu [*]
Department of Chemistry, Faculty of Science, Tokyo University of Science, 1-3 Kagurazaka, Shinjuku-ku Tokyo, Japan

ABSTRACT

It is well known that Weigert effect emerges optical anisotropy induced by linearly polarized light in the case of azo dyes, liquid crystals, and such polymers and its application for holography. We have reported on supramolecular interactions for controlling molecular arrangement of photo-inactive metal complexes between directly photo-functional azo dyes and photo-inactive metal complexes in flexible polymer films. In this work, we have prepared PMMA cast films of three azo dyes, such as azobenzene (**AZ**), disodium 6-hydroxy-5-((2-methoxy-5-methyl-4-sulfophenyl)azo)-2-naphthalenesulfonate (Allura Red), 4,4'-di-n-octyloxyazoxybenzene, and hybrid materials containing a new chiral Schiff base Co(II) complex ($C_{32}H_{32}CoN_2O_2$ showing IR band (C=N) at *c.a.* 1630 cm^{-1}) as a chiral dopant. After circularly polarized UV light irradiation, we compared and found that the degree of chiral ordering (increasing of CD intensity) of **AZ** was the highest among three ones. 4,4'-Di-*n*-octyloxyazoxybenzene having long chain exhibited characteristic increasing of dichroism at π–π* band, which may affect on chiral ordering. With the aid of computational chemistry, we attempted to interpret interactions between transition dipole moments of azo dyes (especially **AZ** was effective) and Co(II) complexes as their light-induced helical arrangement. Theoretical treatment with ZINDO calculations of spectra about the angles of transition dipole moments suggested that supramolecular interactions induced by polarized light predominantly resulted from π–π* transition for **AZ**+PMMA, whereas n-π* transition for Co(II) complex+**AZ**+PMMA.

INTRODUCTION

One of the advantages of organic/inorganic hybrid materials may be easy design for supramolecular multifunctional materials, for example metal complexes incorporating azo-moiety with large dichromic ratios [1]. According to recent development of this field, reversible photo-switching magnetic materials composed of photochromic organic compounds and magnetic inorganic compounds driven by light have been reported [2]. Furthermore, since other molecular features of metal complexes can be used potentially, for example intermolecular fluorescence [3], molecular recognition by chirality [4], single-molecule magnets as well as molecule-based magnets [5], we have employed various metal complexes as the inorganic components of such organic/inorganic hybrid materials.

In this context, we have also focused on photo-tuning of optical anisotropy (molecular alighment) of azo-components by (linearly) polarized UV light irradiation (Weigert effect) [6] in contrast to conventional cis-trans photoisomerization of photocromic azo-compounds. On the other hand, circularly polarized UV light irradiation can induce or control chiral or helical molecular arrangement of such organic/inorganic hybrid materials. Combination of linearly or circularly polarized UV light irradiation and polarized spectroscopy or CD spectroscopy can control or be observed molecular orientation potentially, though intermolecular effects are unclear without chiral fillers or previous chiral filler + chiral complex + PMMA cast film systems [7].

Herein we have investigated three azo-dyes (azobenzene (**AZ**), Allura Red AC (**AR**), and 4,4'-di-n-octyloxyazoxybenzene (**DZ**)) (Figure 1(a)) to select effective photoresponsive components and improved an **AZ** system with the assistance of chiral filler (a chiral Schiff base Co(II) complex (**Co**)) (Figure 1(b)). In this study, we attempted to interpret anisotropic parameters based on transition dipole moments evaluated by ZINDO calculations. In addition, we have assembled a model of interaction by using the dipole moment of the azo dyes calculated to discuss supramolarular chirality induced by polarized UV light.

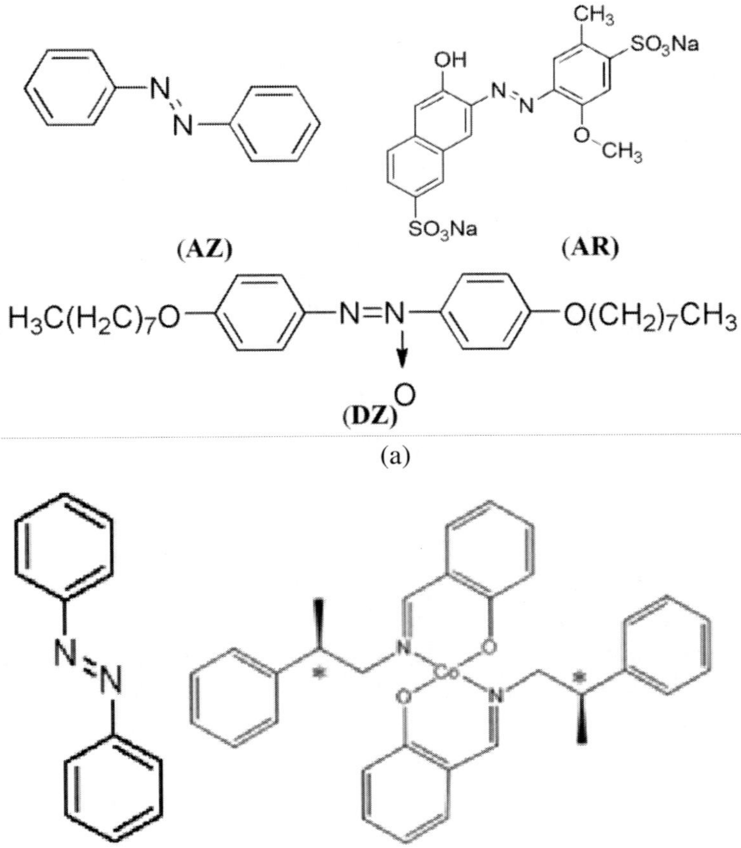

Figure 1. Molecular structures of (a) azo-dyes (**AZ**, **AR**, and **DZ**) and (b) components of hybrid material **AZ+Co**.

EXPERIMENTAL SECTION

Materials

All reagents including three azo dyes (**AZ**, **AR**, and **DZ**) and solvents were commercially available (Wako, TCI, and Aldrich) and were used as purchased without further purification.

Preparations of Bis (N-*R*-2-Phenlypropyl-Salicydenaminato) Cobalt(II) (Co)

To a solution of benzaldehyde (0.610 g, 5.00 mmol) dissolved in methanol (50 mL) and *R*-(+)-β-methylphenythylamine (0.676 g, 5.00 mmol) was added and stirred at 313 K for 2 hr under nitrogen atmosphere to give rise to orange solution of ligand. Then cobalt(II) sulfate heptahydrate (0.703 g, 2.50 mmol) was added and stirred for 4 hr and the resulting green precipitate of product was filtered. Yield 0.2166 g (16.25 %). IR (Nujol mull): 1620 cm^{-1} (C=N).

Preparations of Hybrid Materials

0.5 mL of acetone solution (0.78 mol/L) of azobenzene (**AZ**), disodium 6-hydroxy-5-((2-methoxy-5-methyl-4-sulfophenyl)azo)-2-naphthalenesulfonate (Allura Red) (**AR**), 4,4'-di-n-octyloxyazoxybenzene (**DZ**), or **AZ+Co** (10 mmol) and 1.0 mL of acetone solution of poly(methyl methacrylate (PMMA) were cast onto a slide glass and dried for overnight at room temperature to give rise to **PMMA** film of hybrid **AZ+PMMA**, **AR+PMMA**, **DZ+PMMA**, and **AZ+Co+PMMA**, respectively.

Physical Measurements

Infrared spectra were recorded a JASCO FT-IR 4200 spectrophotometer equipped with a polarizer in the range of 4000-400 cm^{-1} at 298 K. Electronic spectra were measured on a JASCO V-570 spectrophotometer equipped with a polarizer in the range of 900-200 nm at 298 K. Circular dichroism (CD) spectra were measured on a JASCO J-725 spectropolarimeter in the range of 800-200 nm at 298 K. UV and visible light source used was Hayashi LA-310UV and LA-251Xe, respectively with visible ($\lambda > 350$ nm) or UV ($\lambda < 350$ nm) cut filters and (λ/4-) polarizers.

Computational Methods

All calculations were performed using the Gaussian 09W software Revision A.02 (Gaussian, Inc.) [8]. The gas phase geometry optimizations were carried out using ZINDO with B3LYP functional. The vertical excitation energy was calculated with the 6-31+G(d) basis set for H, C, N, and O based on the singlet ground state geometry.

Results and Discussion

Computational Results

Figure 2 indicates calculated direction and magnitude of dipole moments for azo-dyes. Optimized structures were calculated under assuming cis-isomers which produced after UV light irradiation in a PMMA matrix. The evaluated dipole moments are 3.154 Debye, 21.26 Debye, and 5.983 Debye for AZ, **AR**, and **DZ**, respectively.

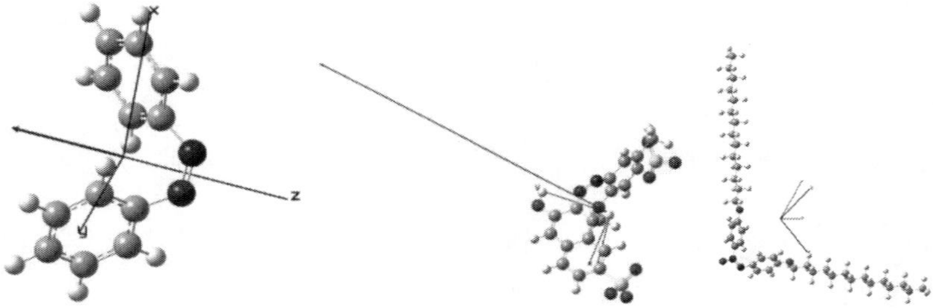

Figure 2. Optimized structures with calculated dipole moment for **AZ** [left], **AR** [middle], and **DZ** [right].

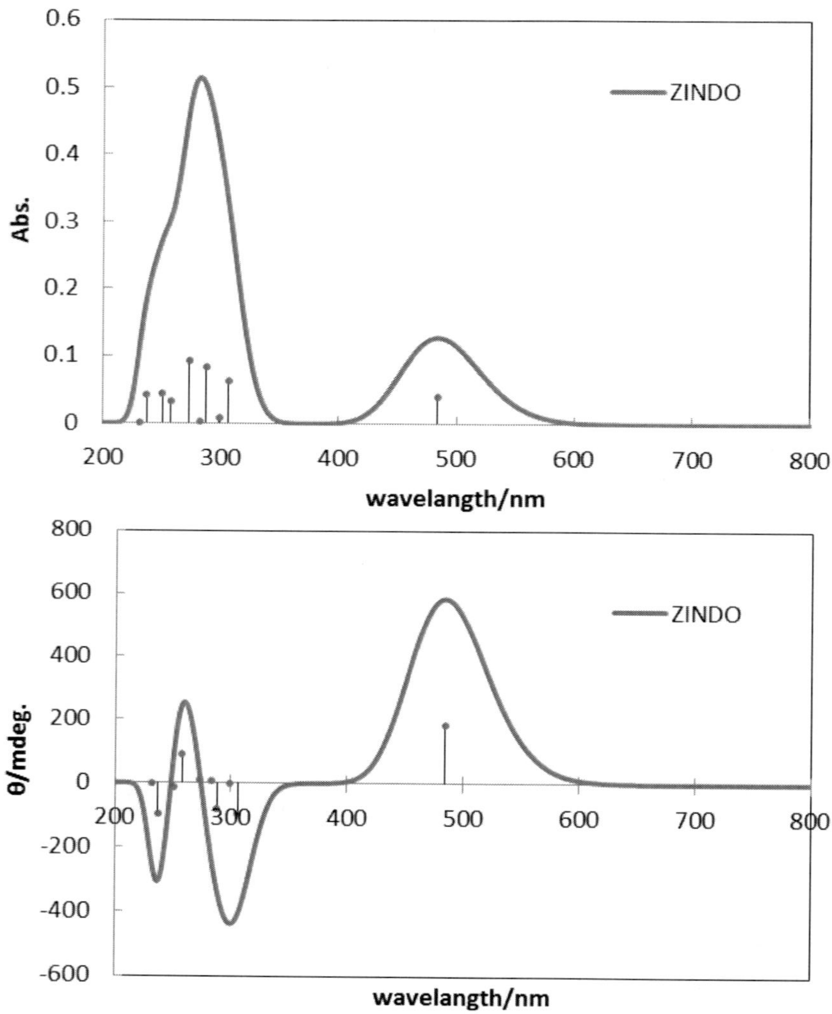

Figure 3. Simulated electronic spectra [above] and estimated (virtual) CD spectra [below] for **AZ**.

Figures 3-5 show simulated absorption spectra (transition energies and their intensity were calculated with ZINDO for the optimized structures and line width was added as Gaussian curves) and 'expected' simulated CD spectra (based on calculated transitions with sign of optical rotation by means of ZINDO and magnetic dipole transitions were assumed for CD) for achiral azo-dyes **AZ**, **AR**, and **DZ**, respectively. It should be noted that actual azo-dyes, **AZ**, **AR**, and **DZ**, were achiral, in other word, did not appear the CD spectra, and (simulated) CD spectra were calculated to discuss assignment of CD bands due to supramolecular chirality in the later section. Because experimental CD spectra is impossible to determine, this is advantage of computational results essentially.

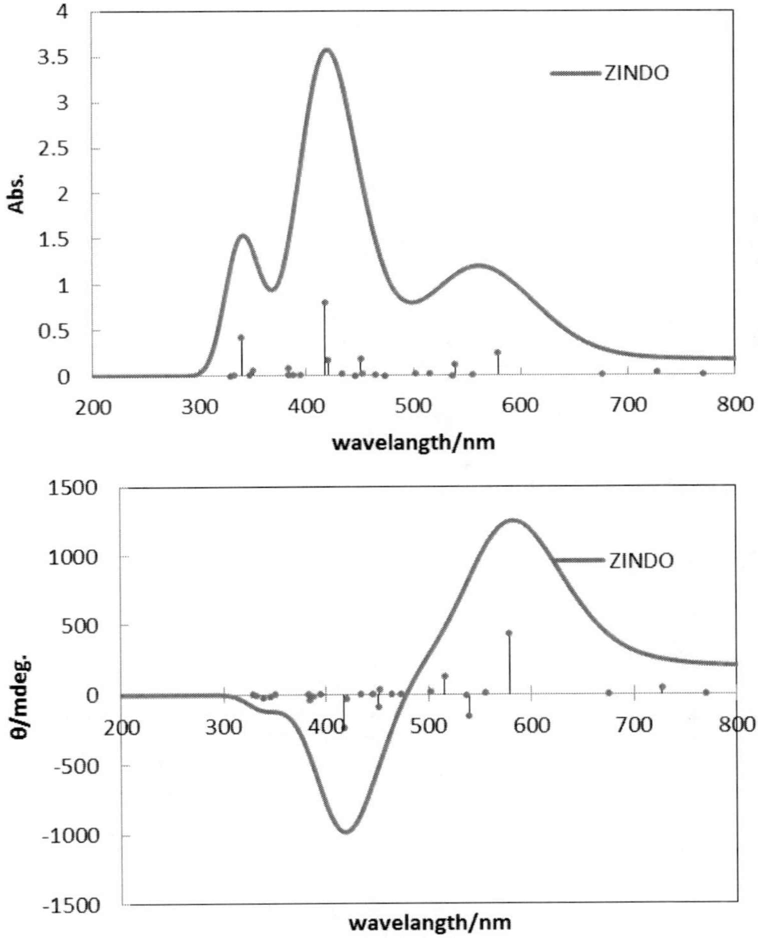

Figure 4. Simulated electronic spectra [above] and estimated (virtual) CD spectra [below] for **AR**.

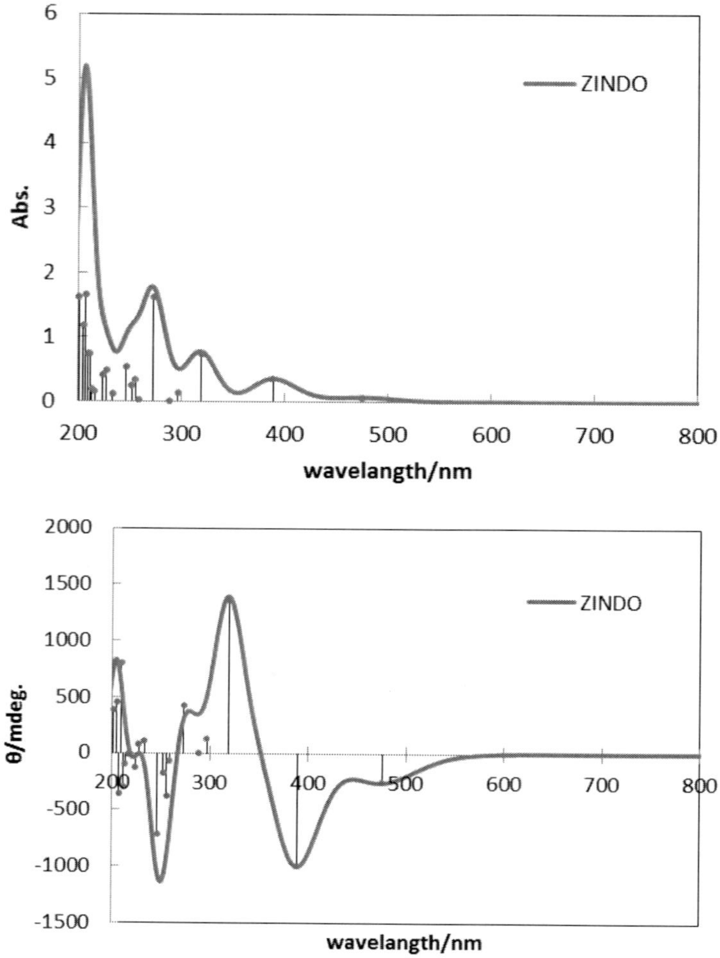

Figure 5. Simulated electronic spectra [above] and estimated (virtual) CD spectra [below] for **DZ**.

Linearly Polarized UV Light-Induced Molecular Arrangement

After linearly polarized UV light irradiation, intensity (absorbance as it measured) of some characteristic peaks of polarized absorption spectra gradually were measured. Figures 6-8 were obtained with polarizer aligned at 0 degree before (0 min) and after liner polarized UV light irradiation for 10 min for hybrid materials **AZ+PMMA**, **AR+PMMMA**, and **DZ+PMMA**, respectively. They were also measured for 0-90 degree at every 5 degree before and after linearly polarized UV light irradiation for 1, 3, 5 and 10 min.

Figure 6 shows polarized absorption spectra and circular diagrams of angular dependence of absorbance of π–π* and n-π* bands at 318 nm and 440 nm for **AZ+PMMA**. It should be noted that large change of π–π* band at initial stage is attributed to photoisomerization from the trans-form of AZ to the cis-form of AZ by UV light irradiation. These circular diagrams about π–π* bands suggest clear optical anisotropy of AZ, which is similar to polymers containing azo-groups [9].

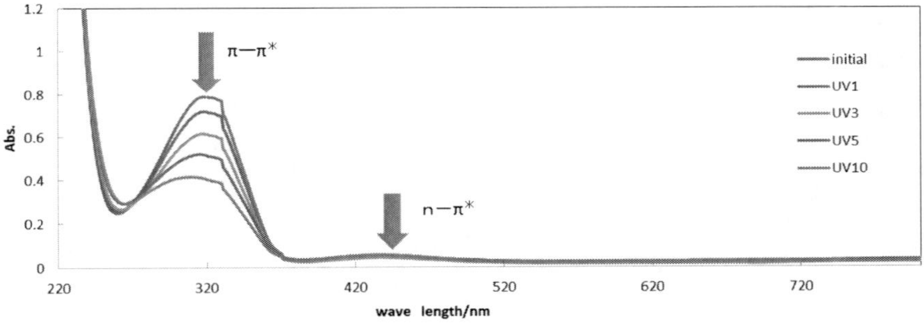

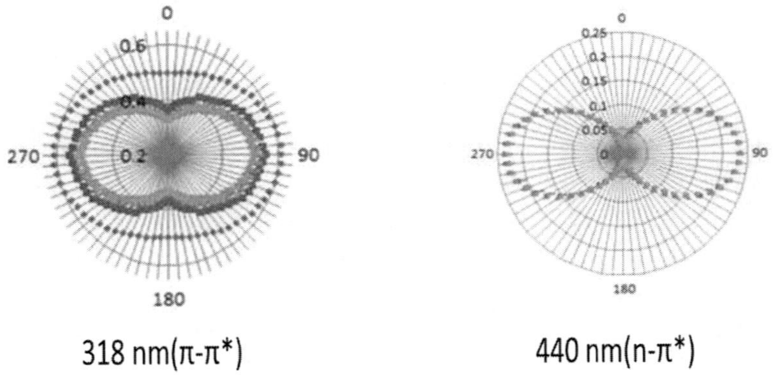

Figure 6. Polarized electronic spectra [above] and circular diagrams [below] for **AZ+PMMA**.

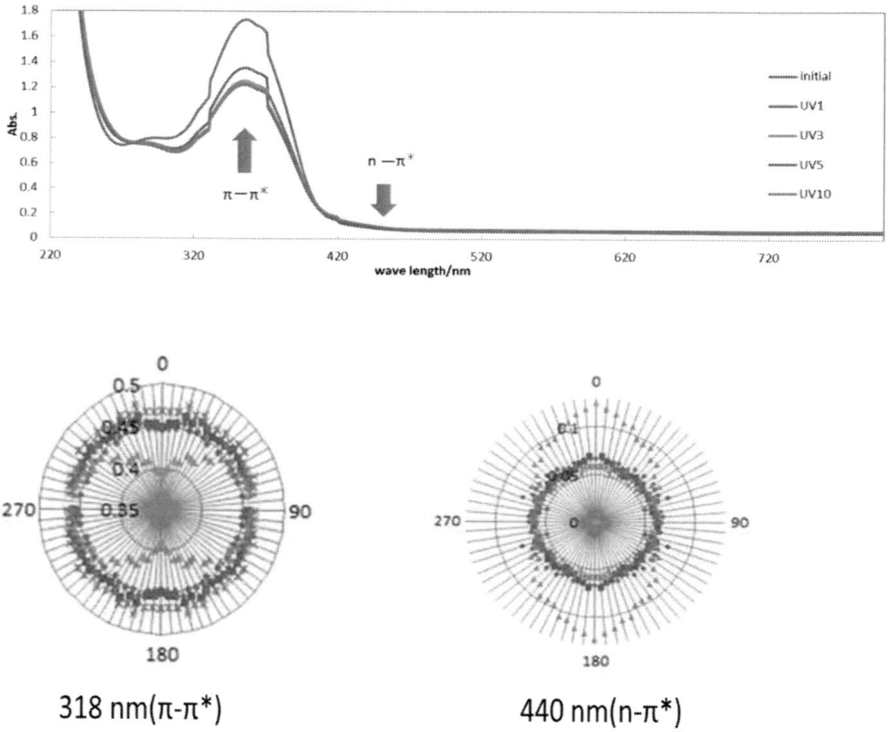

Figure 7. Polarized electronic spectra [above] and circular diagrams [below] for **DR+PMMA**.

Figures 7 and 8 show polarized electronic spectra and circular diagrams of angular dependence of absorbance of π–π* and n–π* at 318 nm, 440 nm, for **AR+PMMA** and **DZ+PMMA**, respectively. It should be noted that only **AZ+PMMA** was influenced clearly by Weigert effect directly among two components. Therefore, **AZ** is the most effectively photo responsible azo-dyes among three samples.

Comparing with π–π* and n–π* bands at 318 nm, 440 nm, respectively, in short-wavelength (UV) region of experimental spectra (Figures 6-8), calculated electronic spectra (Figures 2-4) were possible to reproduce qualitative features of the bands as well as the transitions.

In order for quantitative discussion of optical anisotropy, the S and R parameters were employed. The degree of photoinduced optical anisotropy (spectral dichroism) of polarized absorption electronic spectra can be described commonly by these two parameters:

$$S = (A_{\text{parrallel}} - A_{\text{perpendicular}}) / (2A_{\text{perpendicular}} + A_{\text{parrallel}})$$
$$R = A_{\text{perpendicular}} / A_{\text{parrallel}}$$

where $A_{\text{perpendicular}}$ and $A_{\text{parrallel}}$ values denote absorbance measured with the measuring polarizers perpendicular or parallel to electric vector of irradiation polarized light. Ideal isotropic systems indicate $S = 0$ and $R = 1$ and both S and R parameters are changed as increasing spectral dichroism by molecular alignment.

Table 1 summarized these parameters and revealed that the order of optical anisotropy induced by linearly polarized UV light was **AZ+PMMA > AR+PMMA > DZ+PMMA**.

Table 1. The R and S values after linearly polarized UV light irradiation for AZ+PMMA, AR+PMMA, and DZ+PMMA

AZ+PMMA				
UV time(min)	π-π*(318 nm)		n-π*(440 nm)	
	R	S	R	S
0	0.81216	-0.0668	0.07549	-0.4454
1	0.66003	-0.1278	0.08517	-0.4387
3	0.62809	-0.1415	0.08501	-0.4388
5	0.62288	-0.1438	0.07592	-0.4451
10	0.64778	-0.1330	0.09538	-0.4317
AR+PMMA				
0	1.02075	0.00687	1.00839	0.00279
1	1.03382	0.01115	1.12017	0.03852
3	1.30258	0.09162	2.06155	0.26137
5	0.98848	-0.0039	0.98618	-0.0046
10	0.97111	-0.0097	0.9212	-0.0270
DZ+PMMA				
0	0.99626	-0.00125	0.97223	-0.00934
1	0.99055	-0.00316	1.19129	0.05994
3	0.95376	-0.01566	0.97950	-0.00688
5	0.95569	-0.01499	0.93995	-0.02042
10	0.94791	-0.01767	0.95967	-0.01362

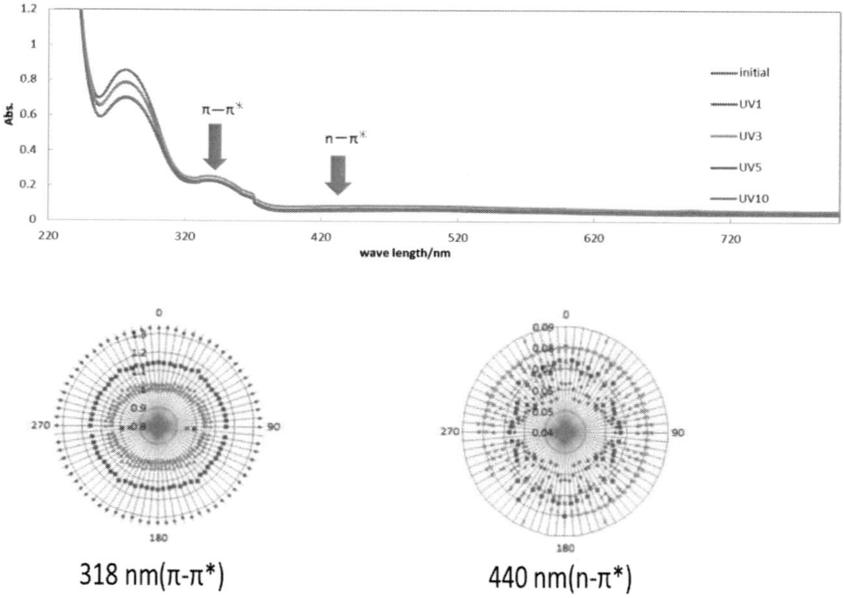

Figure 8. Polarized electronic spectra [above] and circular diagrams [below] for **DZ+PMMA**.

Linearly Polarized UV Light-Induced Molecular Arrangement for AZ+Co+PMMA

As for **AZ** only, the most effective azo-dye, the effect of photo-inactive chiral dopant (**Co**) was also investigated. After linearly polarized UV light irradiation, intensity (absorbance as it measured) of some characteristic peaks for not only **AZ** but also **Co** was measured. Figure 9 shows polarized electronic spectra and circular diagrams of angular dependence of absorbance of π-π* band, n-π* band, and d-d bands due to **Co** at 318 nm, 440 nm, and 520 nm for **AZ+Co+PMMA**. Table 2 summarized the corresponding R and S values. It should be noted that only **AZ** was influenced by Weigert effect directly among two components. Therefore, increasing optical anisotropy detected by the d-d band at 520 nm for **AZ+Co+PMMA** is attributed to **Co** which is also obvious proof of supramolecular interaction between **AZ** and **Co**.

Reducing direct transmission of molecular alignment from **AZ** by flexibility of metal complexes due to Co(II) ions is considered for Schiff base metal complexes [10], discrepancy of tendency about molecular orientation between **AZ+Co+PMMA** cannot be explained reasonably. Therefore, to interpret dichroism of each component properly, dipole-dipole interactions (which could not be estimated from experimental data straightforwardly) between **AZ** and **Co** should be treated appropriately.

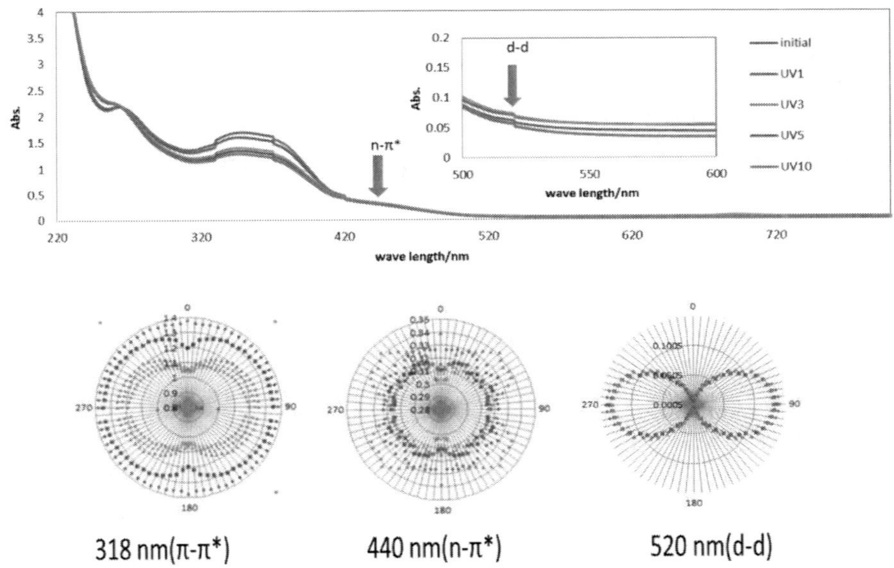

Figure 9. Polarized electronic spectra [above] and circular diagrams [below] for **AZ+Co+PMMA**.

Table 2. The R and S values after linearly polarized UV light irradiation for AZ+Co+PMMA

AZ+Co+PMMA				
UV time(min)	π-π*(318 nm)		n-π*(440 nm)	
	R	S	R	S
0	0.99826	-0.00056	1.04761	0.01562
1	0.88750	-0.03896	0.98047	-0.00655
3	0.84688	-0.05378	0.99987	-0.000042
5	0.85647	-0.05025	0.99556	-0.00148
10	0.86792	-0.04605	0.97422	-0.00867

Circularly Polarized UV Light-Induced Molecular Arrangement

In order to induce chiral (or helical) molecular arrangement by means of circularly polarized UV light, we investigate increasing CD bands (actually difference in intensity of a certain CD peaks) for **AZ+PMMA**, **AR+PMMA**, **DZ+PMMA**, and **AZ+Co+PMMA** (Figures 10-13, respectively). Photo-responsibility may be effective for **AZ+PMMA**, and **AZ+Co+PMMA** should be compared to discuss the effect of chiral dopant. In the linearly polarized light irradiation, **AZ+Co+PMMA** was confirmed optical anisotropy of up to 3 minutes after irradiation. In addition, the circularly polarized light irradiation, **AZ+Co+PMMA** was confirmed after irradiation 10 minutes.

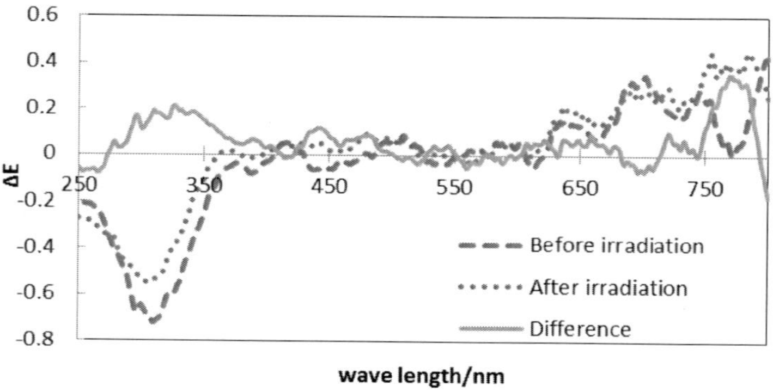

Figure 10. Difference of CD spectra after circularly polarized UV light for 3 min **AZ+PMMA**.

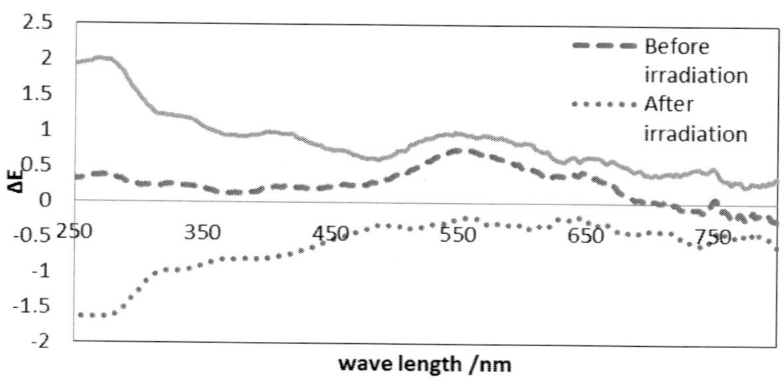

Figure 11. Difference of CD spectra after circularly polarized UV light for 3 min **DR+PMMA**.

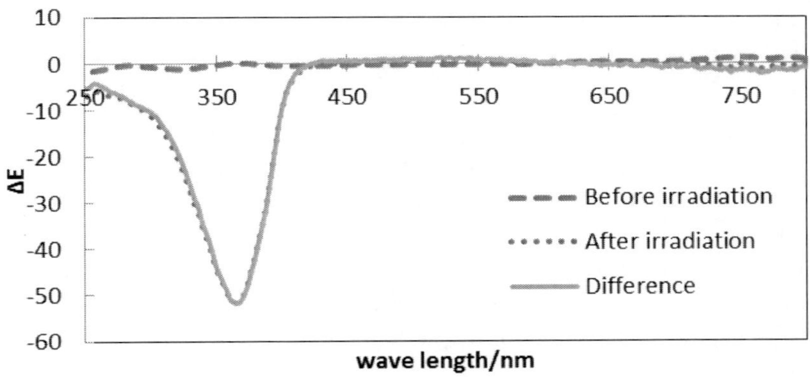

Figure 12. Difference of CD spectra after circularly polarized UV light for 3 min **DR+PMMA**.

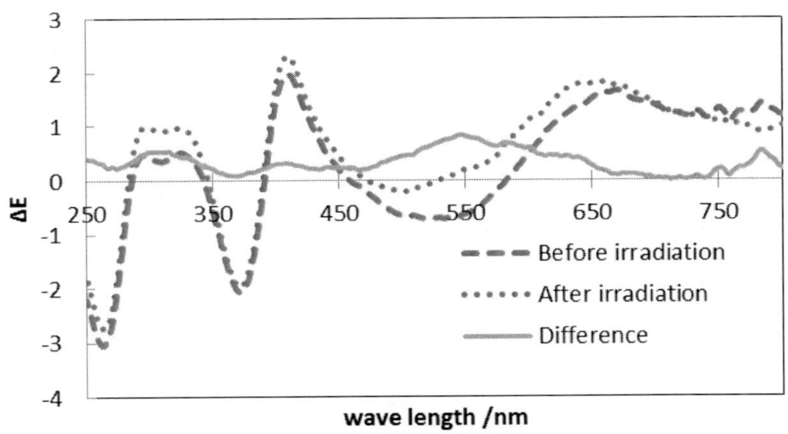

Figure 13. Difference of CD spectra after circularly polarized UV light for 10 min **AZ+Co+PMMA**.

Comparing with experimental results of CD spectra and "theoretical" CD spectra, the increasing bands of CD after circularly polarized UV light irradiation is not attributed to transitions due to molecular transitions because of absence of the corresponding transition. Therefore, emerging CD bands may be attributed to supramolecular chirality due to helical molecular arrangement induced by circularly polarized UV light irradiation. Assistance of the chiral dopant (**Co**) for **AZ** is agreement with this interpretation.

Examination of the Optical Rotation Strength by Coupled Oscillator Model

In order for qualitative interpretation of supramolecular chirality, we attempt to discuss spatial relationship between azo-dye molecules assuming dipole-dipole interaction. For this reason, we will apply coupled oscillator model [11].

If only the Coulomb interaction of the chromospheres A between B two can be considered, we assume system dipole interactions paired the two occurs. Now, the Hamiltonian of the system you are thinking is represented by the equation (1).

$$H = H_A(r_A) + H_B(r_B) + V_{AB} \tag{1}$$

Where Hamiltonian $H_A\,(r_A)$ or $H_B\,(r_B)$ is of the chromophore *A* or *B*, V_{AB} is Coulomb interaction potential between the chromophore *A* and *B*. By the wave function of the ground states ($\Phi_0{}^A$or $\Phi_0{}^B$) and excited states (Φ_a or Φ_b) of the chromophore *A* or *B*, to introduce electric transition dipole moment, optical rotation strength of the formula is derived as the equation (2).

$$R_{0a} = -\left(\frac{2\pi v_a v_b}{hc(v_a{}^2 - v_b{}^2)}\right)(\mu_{0a} \cdot T \cdot \mu_{0b})(R_{AB} \times \mu_{0a} \cdot \mu_{0b}) \tag{2}$$

Where v_a or v_b is absorption frequency, μ_{0a} or μ_{0b} is electric transition dipole moment, R_{AB} is a distance of *A* and *B*, *T* is parameters.

To simplify this, consider that both chromospheres are identical, and hereafter assume R_{AB} connecting both two μ_{0a}, μ_{0b} is orthogonal (Figure14).

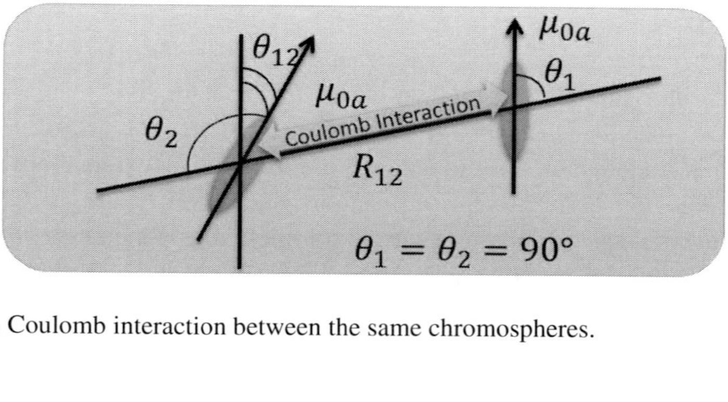

Figure14. Coulomb interaction between the same chromospheres.

At this time, from equation (2), the optical rotation strength $R_{\pm}$ becomes the following equation (3).

$$R_{\pm} = \mp \left({}^{\pi \upsilon_0}/_{2_c} \right) R_{12} \mu_{0a}{}^2 \sin\theta_{12} \tag{3}$$

Further, we considered $R_{\pm}$ as follows. The intensity of CD bands can be represented by optical rotation strength *R*. From the CD spectra measured, *R* is determined according to the following equation (4).

$$R = 2.296 \text{ x } 10^{-39} \int_0^{\infty} \left\{ {}^{\Delta\, \varepsilon(\sigma)}/_{\sigma} \right\} d\sigma$$
$$\approx \left({}^{2.296 \text{ x } 10^{-39}}/_{\sigma_0} \right) \int_0^{\infty} \Delta\varepsilon(\sigma) d\sigma \text{ (c.g.s)} \tag{4}$$

where σ is the wavenumber and σ_0 is the wavenumber of the peaks (the maximum) of the CD spectra.

As relational expressions, we calculated from equation (4) optical rotation strength of the induced CD spectra observed (see Figures 10-13). Substituting equation (3), the degree of inclination (θ) of the dye molecules of interest was evaluated. The transition dipole moment at this time is based on the results of ZINDO. As a result, the angle of optical rotation and strength to transition electric dipole moment of each dye in the π-π* or n-π* transition state of each systems was summarized in Table 3.

Table 3. The angle of optical rotation and strength to transition electric dipole moment of each dye in the *π*-*π or n-*π** transition states. For simplicity, R_{12} was assumed to be 0.5 nm all. In AZ+Co+PMMA, Co is considered as a chiral dopant**

	AZ+Co+ PMMA	AZ+ PMMA	AR+ PMMA	DZ+ PMMA
R (n-π*)/c.g.s	6.25529×10^{-34}	1.03224×10^{-34}	1.69077×10^{-34}	4.20159×10^{-33}
θ / deg	3.490030×10^{-29}	5.75921×10^{-30}	9.59665×10^{-35}	4.22290×10^{-36}
$\mu_{0a}(n\text{-}\pi^*)$/*Debye*	0.795802	0.795803	0.023932	0.587770
R (π-π*)/c.g.s	3.95981×10^{-34}	6.53445×10^{-35}	1.24778×10^{-34}	2.81683×10^{-33}
θ/deg	1.40474×10^{-29}	2.31809×10^{-30}	1.95930×10^{-34}	2.69171×10^{-37}
$\mu_{0a}(\pi\text{-}\pi^*)$/*Debye*	0.794052	0.794052	0.012360	1.560796

The calculated results indicated that transition dipoles of **AZ** resulting in large electric dipole transition of n-π* and π-π* transitions are significantly inclined each other, which was attributed to mutual molecular orientation induced by circularly polarized UV light irradiation. In addition, comparison between A**Z+Co+PMMA** and **AZ+PMMA** suggested that chiral dopant (**Co**) strongly assisted to chiral molecular orientation of **AZ** induced because of the Coulomb interaction between **Co** and **AZ**.

CONCLUSION

Consequently, similar to linearly polarized UV light exhibited Weigert effect, circulary polarized UV light could induce chiral supramolecular arrangement of azo-dyes, **AZ**, **DR**, and **DZ**, and **AZ** is the most effective photosensitive azo-dye among them due to suitable to flexibility in PMMA environment. Additionally, chiral dopant (**Co**) could gain the effect to form chiral supramolecular arrangement of **AZ** after circulary polarized UV light radation, which was led by model discussion of dipole-dipole interaction to interpret n-π* and π-π* transitions.

ACKNOWLEDGMENT

This work was supported by the Research Foundation for Opto-Science and Technology.

REFERENCES

[1] Blackburn, O. A., Coe, B. J., Fielden, J., Helliwell, M., McDounall, J. J. W. & Hutchings, M. G. (2010). *Inorg. Chem.*, *49*, 9136 and references therein; A. A. Khandar and Z. Revani, Polyhedron, 1998, 18, 129.

[2] Einaga, Y., Taguchi, M., Li, G., Akitsu, T., Sugai, T. & Sato, O. (2003). *Chem. Mater.*, *15*, 8, R. Mikami, M. Taguchi, K. Yamada, K. Suzuki, O. Sato and Y. Einaga, Angew. *Chem. Int.* Ed., *43*, 6135, M. Taguchi, K. Yamada, K. Suzuki, O. Sato and Y. Einaga, *Chem. Mater.*, 2003, *17*, 4554., T. Yamamoto, Y. Umemura, O. Sato and Y. Einaga, *J. Am. Chem. Soc.*, 2005, *127*, 16065, T. Yamamoto, Y. Umemura, O. Sato and

Y. Einaga, *Chem. Mater.*, 2004, *16*, 1195, Y. Einaga, R. Mikami, T. Akitsu and G. Li, *Thin Solid Films*, 2005, 493, 230.

[3] Akitsu, T. & Yoshida, A. (2011). *Curr. Phys. Chem.*, *1*, 76, T. Akitsu and S. Yamamoto, *Asian Chem. Lett.*, 2010, 14, 255.

[4] Akitsu, T. & Einaga, Y. (2005). *Polyhedron*, *24*, 1869, Akitsu, T. & Y. Einaga, Polyhedron, 2005, 24, 2933, T. Akitsu, Polyhedron, 2007, 26, 2527.

[5] Akitsu, T. & Einaga, Y. (2006). *Inorg. Chem. Commun.*, *9*, 1108, Akitsu, T. & J. Nishijo, J. Magn. Magn. Matter., 2008, 320, 1586, Akitsu, T. & J. Nishijo, J. Magn. Magn. Matter., 2007, 315, 95, T. Akitsu, J. Magn. Magn. Matter., 2009, 321, 207.

[6] Weigert, F. (1921). *Naturwissenschaften*, *29*, 583, K. Ichimura, Chem. Rev., 2000, 100, 1847 and references therein.

[7] Akitsu, T., Ishioka, C. & Itoh, T. (2009). *Cent. Euro. J. Chem.*, *7*, 690, Akitsu, T. & C. Ishioka, Asian Chem. Lett., 2010, 14, 37, Akitsu, T. & T. Itoh, Polyhedron, 2010, 29, 477, Akitsu, T. & R. Tanaka, Curr. Phys. Chem., 2011, 1, 82, Akitsu, T. & Y. Miura, J. Chem. Chem Eng.,2011, 5, 443, Y. Aritake, T. Takanashi, A. Yamazaki and T. Akitsu Polyhedron, 2011, 30, 886.

[8] Frisch, M. J., Trucks, G. W., Schlegel, H. B., Scuseria, G. E., Robb, M.A., Cheeseman, J. R., Scalmani, G., Barone, V., Mennucci, B., Petersson, G. A., Nakatsuji, H., Caricato, M., Li, X., Hratchian, H. P., Izmaylov, A. F., Bloino, J., Zheng, G., Sonnenberg, J. L., Hada, M., Ehara, M., Toyota, K., Fukuda, R., Hasegawa, J., Ishida, M., Nakajima, T., Honda, Y., Kitao, O., Nakai, H., Vreven, T., Montgomery, J. A. Jr., Peralta, J. E., Ogliaro, F., Bearpark, M., Heyd, J. J., Brothers, E., Kudin, K. N., Staroverov, V. N., Kobayashi, R., Normand, J., Raghavachari, K., Rendell, A., Burant, J. C., Iyengar, S. S., Tomasi, J., Cossi, M., Rega, N., Millam, J. M., Klene, M., Knox, J. E., Cross, J. B., Bakken, V., Adamo, C., Jaramillo, J., Gomperts, R., Stratmann, R. E., Yazyev, O., Austin, A. J., Cammi, R., Pomelli, C., Ochterski, J., Martin, R. L., Morokuma, K., Zakrzewski, V. G., Voth, G. A., Salvador, P., Dannenberg, J. J., Dapprich, S., Daniels, A. D., Farkas, O., Foresman, J. B., Ortiz, J. V., Cioslowski, J. & Fox, D. J. (2009). GAUSSIAN 09 (Revision A.1), Gaussian, Inc., Wallingford, CT.

[9] Li, X., Lu, X., Lu, Q. & Yan, D. (2007). *Macromolecules*, *40*, 3306, B. Sapich, A. B. E. Vix, J. P. Rabe and J. Stumpe, Macromolecules, 2005, 38, 10480, R. Advincula, M.-K. Park, A. Baba and F. Kaneko, Langmuir, 2003, 19, 654, T. Geue, A. Ziegler and J. Stumpe,

Macromolecules, 1997, 30, 5729, N. Bohm, A. Materny, W. Kiefer, H. Steins, M. M. Muller and G. Schottner, Macromolecules, 1996, 29, 2599, N. Bohm, A. Materny, H. Steins, M. M. Muller and G. Schottner, Macromolecules, 1998, 31, 4265, L. De Boni, C. Toro, A. E. Masunov and F. E. Hernandez, J. Phys. Chem. A, 2008, 112, 3886.

[10] Sakiyama, H., Okawa, H., Matsumoto, N., Kida, S. J. (1990). *Chem. Soc., Dalton Trans.*, 2935, B. Bosnich, J. Am. Chem. Soc., 1968, 90, 627. H. Okawa, M. Nakamura and S. Kida, Inorg. Chim. Acta, 1986, 120, 185, Y. Nishida and S. Kida, Bull. Chem. Soc. Jpn., 1970, 43, 3814, H. Sakiyama, H. Okawa, N. Matsumoto and S. Kida, Bull. Chem. Soc. Jpn. 1991, 64, 2644. M. Ulusoy, H. Karabiyik, R. Kilincarslan, M. Aygun, B. Cetinkaya and S. Garcia-Granda, Struct. Chem., 2008, 19, 749, Akitsu, T. & Y. Einaga, Polyhedron, 2006, 25, 1089.

[11] Kanbara, S. & Nanzyo, M. (1984). "Spectroscopy in the polymer", Kyorotsu Publication, Japan, C. L. Cech, W. Hug, I. Tinoco Jr., Biopolymers, 15, 131(1976)., D.J. Caldwell, H.Eyring, "Theory of Optical Activity", Wiley-Interscience, New York (1971)., I. Tinoco, Jr., *Advan. Chem. Phys.*, *4*, 113(1962).

In: Computational Chemistry
Editor: Daria Bove
ISBN: 978-1-63321-354-8

Chapter 6

A General Procedure for a Priori Calculation of Thermochemical Properties of Organic Molecules and Free Radicals

Arijit Bhattacharya[1,*], Yuvraj Dangat[2] and Kumar Vanka[2]

[1]Chemical Engineering Division, National Chemical Laboratory, Pune
[2]Physical Chemistry Division, National Chemical Laboratory, Pune

Abstract

There have been many attempts in the literature to use variations of the "3rd law" method for estimating several thermodynamic properties (including the entropy and the heat capacity) for a variety of molecules. Application of the method to cases of free radicals is much less common. This is because of the difficulties and uncertainties associated with the assignment and the calculation of the parameters required for providing the higher order anharmonicity related corrections to improve upon the values of the properties obtained under the basic rigid-rotor harmonic-oscillator (RRHO) approximation.

---

* Corresponding author email: arijit.sb@gmail.com. Current Address: G-203, Pinnac Kanchanganga Co-operative Housing Society (Opp. CONVERGYS), Aundh, Pune 411007.

The study reported in this chapter devised and standardised a procedure based on the consistent theoretical premises to make a priori estimates of ideal gas entropy and heat capacity of organic molecules and especially free radicals with acceptable accuracies. An important part of the procedure is to calculate in a transparent and scalable manner the higher order corrections due to hindered rotation (HR) of groups internal to the molecular species to be added to their RRHO approximated property values. Our HR correction procedure adapts the classical Pitzer method to a range of large and branched radicals containing multiple rotating tops, either symmetric or asymmetric, with single or repeat occurrences of a given top within the same radical.

Although a commercially available software (such as Gaussian) may allow a user an automatic way to calculate the corrections of the property values due to hindered rotations (HR) at some low vibrational frequencies, our experience suggests that caution should be exercised before accepting the software output in this regard. The corrected property values for a number of different classes of radicals, published by previous researchers, were found to be more closely reproduced by using the methodology presented in this chapter than using Gaussian software's HR correction facility.

For larger molecules and radicals that may exist as multiple low-energy stable conformers, an additional correction due to the presence of a mixture of conformers was used in an approximate manner for entropy only. This correction is expected to make the prediction of entropy more accurate. Gaussian calculation does not seem to account for the conformers.

INTRODUCTION

Energetics and kinetics of many industrially important classes of reactions such as combustion [1], oxidation [2], polymerization as well as the photochemical processes in atmospheric pollution chemistry [3], in biology and medicine [4] require the knowledge of thermodynamic properties of organic free radicals considered to be key reactants and products of the elementary steps constituting the mechanism of the overall reaction. Accurate predictive calculation of thermo-chemical properties (standard state enthalpy of formation, entropy and heat capacity) of radicals by making judicious use of the available modern computational chemistry tools is a capability valued by chemists and chemical engineers.

There is a well-established methodology of making a priori calculation of entropy and heat capacity of molecules in the gas phase by using relations of statistical thermodynamics [5] that was extended to free radicals by Benson and his coworkers many years ago [6]. This so-called "3rd law" method, though simple enough to apply, faced several infirmities about the correct structural parameters, vibration frequencies, molecular symmetries and barrier heights for internal rotations. With the advent of powerful quantum chemistry software for making ab initio calculation of molecular structures and vibrational frequencies, it is now possible to use these molecular property data (along with a correctly assigned radical symmetry) to calculate the ideal gas entropy and heat capacity of radicals to a good first approximation on the basis of the so-called rigid rotor harmonic oscillator (RRHO) assumption.

One of the higher order approximations is consideration of the anharmonicity arising out of the deviation in the spectroscopic energy levels from the pattern assumed within the RRHO model [5] which is often ignored being numerically quite small. The treatment of low vibrational modes as anharmonic vibrations is an active research area and has been accounted for in individual cases [7-10], yet no general procedure is available to date for the treatment of anharmonicity in large molecules [11].

What cannot be disregarded however, especially at higher temperatures, is the correction of the RRHO estimates of the properties due to the contribution of the hindered rotation (with a potential barrier against the rotation not too small nor too large in comparison with kT) of groups of atoms connected by single electron-pair bonds internal to the radical [5]. The seminal work of Pitzer and Gwinn [12] provided a general treatment of the internal rotation of a single symmetric rotating top attached to the remaining rigid part of a molecule. The treatment was subsequently somewhat generalised in an approximate manner by Pitzer [13], and Pitzer and Kilpatrick [14] to cases of non-symmetric tops and to molecules with more than one top. The application of the Pitzer and Gwinn [12] treatment of internal rotation of single rotors has been extended in later studies [15-22] not only to a large number of hydrocarbon molecules, some simple alkyl, haloalkyl radicals but to more complex alkyl peroxy radicals. These cases showed instances of single or multiple symmetric or non-symmetric tops occurring within the same molecule/radical.

The applicability of the classical Pitzer and Gwinn approach [12] to molecular species involving simultaneous occurrence of different rotors or multiple occurrence of the same rotor has been questioned on theoretical grounds and generally believed to be not quite justified. There has been a significant parallel effort in deriving the hindered rotation correction by a direct quantum chemical treatment and called the 1D hindered rotor (HR) approximation. This body of work has been summarized and reviewed by Pfaendtner et al. [11] and will not be pursued further in this chapter. Despite these developments, the above cited attempts at extending the application of the classical Pitzer methodology is generally considered a valid approach leading to consistent and practically acceptable approximate evaluation of the thermo-chemical properties at a much less computational effort.

Possibly proceeding on such a premise a standard and widely used quantum chemistry software, Gaussian [23a,b], include a provision to add, within its thermochemistry calculations, hindered rotation correction to the RRHO property values using an adaptation of the Pitzer-Gwinn methodology due to Ayala and Schlegel [24]. This methodology works without any intervention of the users and can be used with the 'freq = hinderedrotor' keyword. This approach provides an approximate treatment for all possible internal motions decided within the software as qualifying for the HR correction and may not always lead to entirely satisfactory results. That happened to be the experience in our initial experimentation with the software for both the cases of molecules and radicals for which property values have been published in the literature and also for a series of larger and branched radicals consisting of multiple rotating tops that we were investigating as a part of our effort to standardise a consistent methodology to make reliable and accurate estimate of the ideal gas entropy and heat capacity.

In this chapter it has been shown how one can pick up some of the parameters (required for our RRHO and the HR calculations) from the structure optimization and frequency analysis run of a standard quantum chemistry software and then plug those data to a calculation procedure based on the "3rd law" methodology incorporating the standard Pitzer-Gwinn approach [12]. However separate transition state calculations using the same software were used for accurate calculation of the barrier heights.

Finally as suggested by previous workers [22, 25], an entropy of mixing correction was calculated by a simple approximation and added to the RRHO (or the HR-corrected) value for entropy in the case of the existence of a molecule or a radical in more than one low energy accessible conformers.

Our aim would be to first standardise a reliable and transparent calculation procedure incorporating the above elements and validate this methodology by comparing the property values for a variety of free radicals and molecules obtained here with those published in course of similar prior studies in the literature. Finally the methodology will be applied to predict the entropy and heat capacity of a reasonably large set of straight-chain and branched alkyl radicals for which similar predictive studies are not available.

THEORETICAL BACKGROUND

The present study is an attempt to calculate, in a generally usable manner, consistent thermo chemical properties like entropy and heat capacity for organic molecules and free radicals. These properties have been calculated with a first approximation of rigid rotor harmonic oscillator (RRHO) model, with an improvement of the resulting values with a higher order approximation accounting for the hindered internal rotations (HR). The occurrence of multiple molecular conformations when these are present has been taken into account in an approximate manner only for entropy by way of an entropy of mixing correction to the RRHO values. The approximations employed are discussed in detail below.

First Approximation (RRHO)

The contributions to the entropy and heat capacity have been calculated separately for the translational, rotational, vibrational and electronic contributions. The RRHO entropy (S) and heat capacity (C_p) are the overall sum of all the contributions, as shown in the Equations (1) & (2) below.

$$S^o_{(RRHO)} = S^o_{(tr)} + S^o_{(rot)} + S^o_{(vib)} + S^o_{(ele)} \qquad (1)$$

$$C^o_{p(RRHO)} = C^o_{p(tr)} + C^o_{p(rot)} + C^o_{p(vib)} + C^o_{p(elec)} \qquad (2)$$

where subscripts *tr*, *rot*, *vib* and *ele* refer, respectively, to the translational, rotational (external), vibrational and the electronic contributions to the properties. Each of the contributions is calculated by using mathematical expressions available in the standard textbooks and monographs [5, 26-27], which we have assembled here for the sake of self-contained description of the complete procedure used in this chapter.

The *translational* contribution was calculated according to following equations:

$$S^o_{tr} = R\left(\frac{3}{2}\ln M + \frac{5}{2}\ln T\right) - 2.315 \qquad (3)$$

$$C^{o}_{p(tr)} = \frac{5}{2}R \tag{4}$$

where M is the molecular weight of the molecule/radical, T the temperature and R the universal gas constant.

The external rotational contribution to entropy was calculated as:

$$S^{o}_{rot} = R\left[\frac{1}{2}\ln(D\times 10^{117}) + \frac{3}{2}\ln T - \ln\sigma\right] - 0.033 \tag{5}$$

Here, σ is the external symmetry number and D is a product of the three principal moments of inertia, which can be evaluated from the molecule/radical geometry. The heat capacity contribution due to the external rotation for non-linear molecular species (with which we will be concerned in this chapter) was calculated employing the expression below:

$$C^{o}_{p(rot)} = \frac{3}{2}R \tag{6}$$

To calculate the vibrational contributions the vibration analysis data will be required as the input for the following expressions for entropy and heat capacity:

$$S^{o}_{vib} = \sum_{j} R\left[\frac{u_j}{e^{u_j - 1}} - \ln(1 - e^{-u_j})\right] \tag{7}$$

$$C^{o}_{p(vib)} = \sum_{j} R\frac{u_j^2 e^{u_j}}{\left(e^{u_j} - 1\right)^2} \tag{8}$$

where, $u_j = hcw_j / kT$ and w_j is the jth fundamental vibration frequency, h the Planck's constant and k the Boltzmann's constant. Please note that, here, the summation has been taken over all the vibrational modes.

The electronic contribution to entropy is zero if the spin multiplicity (as for a molecule) S=0. However, in the case of the mono-radical, the spin multiplicity is S=1/2. The electronic contribution to the thermodynamic properties is given by the following expressions:

$$S^{o}_{ele} = R \ln Q_{ele} \tag{9}$$

$$Q_{ele} = 2S + 1 \tag{10}$$

where Q_{ele} is the electronic partition function and S is the total electronic spin. The electronic contribution to the heat capacity is zero.

The required inputs to these equations such as the principal moments of inertia and the frequencies of vibrational modes were generated by making a standard run with a software such as GAUSSIAN optimizing the geometry of the molecule or the radical along with the frequency analysis calculation. It is also to be noted that the exact symmetry number is necessary for the calculation of the external rotational contribution to the entropy. In this study, we have assigned the symmetry number generally based on the rules and guidelines given in the literature [28-29]. Sometimes, the output from the software does not report the correct symmetry number and this can lead to the incorrect contribution to the entropy. The importance of correct assignment of the symmetry number in respect of calculations performed with the Gaussian software was pointed out by Guthrie [25]

Using the Equations (1) to (10), the RRHO approximated thermo-chemical properties were calculated and found to be in excellent agreement with the values obtained from the Gaussian 09 [23a] output, provided the external symmetry number was considered properly. These property values have been improved further with the higher approximations that are discussed in the next sub-section.

Higher Approximation: Hindered Internal Rotation

Contributions to entropy and heat capacity due to hindered internal rotation of a rotating group (variously called a rotor or a rotating top, etc.) around a single electron-pair bond attaching the top to the rest of the molecule can be obtained, following Pitzer and Gwinn [12], from the partition function and the potential barrier restricting the rotation. First the partition function for free rotation of the top, $Q_{f(\text{int\_}rot)}$, is calculated from the reduced moment of inertia (I_r) and the symmetry number (σ_{in}) of the rotation. The "free rotor" entropy is then calculated using the following expression:

$$S^o_{f(\text{int\_}rot)} = R\left(\frac{1}{2} + \ln Q_{f(\text{int\_}rot)}\right) \tag{11}$$

with "free rotor" partition function calculated as

$$Q_{f(\text{int\_}rot)} = \frac{1}{\sigma_{in}}\left(\frac{8\pi^3 I_r kT}{h^2}\right)^{1/2} \tag{12}$$

Approximating the potential by a simple cosine function as given by

$$V = V_o\left(1 - \cos(\sigma_{in}\phi)\right)/2 \tag{13}$$

where ϕ is the torsion angle and the height of the potential barrier is V_o, Pitzer and Gwinn [12] had solved the quantum-mechanical problem of the one dimensional rotation subject to this potential and obtained the permitted energy levels that involved complicated expressions, but they could present the final results, namely, the contribution of the internal rotation to entropy, heat capacity along with all other thermodynamic properties as tables in terms of two variables, (a) the ratio of the potential barrier to the thermal energy, V_o/RT, and (b) the inverse of the "free rotor" partition function, $\left(Q_{f(\text{int\_}rot)}\right)^{-1}$. The correction to the entropy is calculated as the entropy loss due to hindered rotation using the Pitzer-Gwinn tables [12] as $\left(S^o_f - S^o_{hr}\right)_{correction}$. The heat capacity correction due to internal rotation is obtained directly from the tables.

It is important to note that the above treatment was originally developed with the assumption that the moments of inertia of the overall rotation of the molecule are independent of the internal rotations i.e., a molecule comprising a rigid frame with attached symmetric tops. Secondly, the tables referred to above can be considered strictly valid for molecules with *one* internal rotational motion and for a potential energy specified as above. In the same paper [12] and a series of subsequent papers Pitzer and his coworkers [13-14] had shown the applicability of the tables to provide good approximations to the hindered rotation corrections in the cases of molecules possessing more than one internal rotational coordinates and potential energy having different shapes and being represented by more general functional forms. The original analysis [12] was extended [13] to asymmetric tops as well. Pitzer and his coworkers further generalised their approximate treatment [14] to cases where rotating groups are attached to other rotating groups, with no limit on the symmetry of the groups or the number of the groups involved, by calculating the kinetic energy matrix for rotation of the molecule, converting the same into a matrix of the internal rotations. The pertinent reduced moment of inertia obtained from this latter matrix allowed one to use the original tables of Pitzer and Gwinn [12] to evaluate the hindered rotation corrections for multiple rotors.

The above theoretical framework has several known limitations e.g., evaluation of the partition function at the classical limit, the nature and the representation of the potential function, largely uncoupled rotors that may be treated independent of each other, etc. Despite that several subsequent researchers have attempted to evaluate the HR corrections for molecules like long chain and branched hydrocarbons [21-22] with repeat occurrence of the methyl top, for haloalkyl radicals containing single (symmetric and asymmetric) tops [15-18] or more complex ones like simple or branched alkyl radicals (e.g., isopropyl and tert-butyl) and the corresponding alkyl peroxy radicals (showing the presence of either multiple repeated symmetric tops or both symmetric and asymmetric tops on the same radical) [19-20].

In a detailed comparison of some of the common methodologies available in the literature [30-31] for approximating the true partition function describing the hindered rotor lying between that of the harmonic oscillator at one end and free rotor at the other, vis-à-vis the Pitzer and Gwinn [12] methodology, Ayala and Schlegel [24] found these as useful approximations that perform fairly well within specified ranges of applicability. The Truhlar formula [30] provides approximation to the free rotor end, but performs poorly as one moves towards hindered rotation. The methodology due to McClurg et al. [31] provides a good approximation for the cases where quantum mechanical partition function at low temperature asymptotically approaches the classical partition function at moderate to high temperature. On making a quantitative comparison, Ayala and Schlegel [24] found that the Pitzer and Gwinn [12] methodology provides the most acceptable approximation to the hindered rotor partition function and thereby the thermodynamic properties within most of the ranges of the variables, namely, the free rotor partition function $Q_{f(\text{int\_}rot)}$ and the reduced barrier height V_o/RT. The performance degrades a little for the low values of the barrier height and Q_f for which they devised a slightly modified formula to take care of the entire range in a uniformly accurate manner.

Ayala and Schlegel [24] provided a complete methodology in their paper to automatically identify the low vibrational frequency modes associated with internal rotation, assign of the symmetry number for the rotation and calculate the relevant barrier height and the reduced moment of inertia to be used in their modified Pitzer-Gwinn formula for calculating the HR correction. The approximation, within its stated limitations, was also extended to the multiple rotor situations.

We concluded, therefore, that despite apparent theoretical limitations of the Pitzer and Gwinn [12-14] method of HR correction discussed in the literature [11, 24] it is a valid approach and could be adapted within our overall methodology to calculate the thermochemical properties of wider range of free radicals than usually attempted.

OCCURRENCE OF MULTIPLE CONFORMATIONS: ENTROPY OF MIXING CORRECTION

The above methodology calculates thermochemical properties like entropy and heat capacity of organic molecules based on a stable minimum energy structural configuration (the so-called *anti*-conformation), with a hindered rotation correction if required.

In reality, however, many of these compounds are known to exist as a mixture of conformations. For instance, butane would be a mixture of an *anti*- with two gauche conformers and it is expected that the properties of such a mixture would be a little different (more or less) than a pure *anti-* form of butane calculated using the method described above. In the context of the theoretical calculation of these properties, therefore, it has been recognised [21, 25] that the contributions from the other conformers have to be accommodated appropriately within the overall calculation methodology.

De Tar [21-22] devised an elaborate procedure which included a strategy for identifying the eligible conformers, calculation of the entropy and heat capacity (also other properties like heat content) of individual conformers and then evaluation of the correction (to be added to the RRHO value) by using an average of the deviations of the property of the conformers from the value for the reference conformer (say, the *anti-* conformer) and adding a mixing term (called the 'entropy of mixing' in the case of entropy). This procedure was applied by the author successfully to a series of hydrocarbons. Guthrie [25] provided a much simpler version of the procedure for entropy where conformers were assumed to have approximately same property values, which reduced the correction to entropy of mixing only. He had applied his method with success to evaluate the entropy of a large number of organic molecules going up to about 10 carbon atoms (and not restricting to hydrocarbons alone). There does not seem to have been any attempt in the literature to correct the "3rd Law" thermochemical properties of free radicals due to presence of conformers.

One is not quite certain if the above methodologies devised for molecules may be easily translated to the cases of radicals with which we are mainly concerned in the present chapter.

In general, there are issues about identifying correct conformations in the case of radicals and choosing the *anti*-conformation as the reference conformer if only because that may not always be the lowest energy conformer. In this work, we have proposed a procedure to approximately calculate the entropy of mixing correction (as the sole contribution due to the presence of a mixture of conformers) and used this in the calculation of entropy of a series of straight chain and branched alkyl radicals. No such correction was attempted for the heat capacity.

A correction to RRHO entropy value due to the presence of more than one low energy conformer was given by Guthrie [25] with the following approximation:

$$S_{mix} = -R\sum_{i=1}^{n} x_i \ln(x_i) \cong R\ln(n) \tag{14}$$

where

$$x_i = \frac{c_i}{\sum_{i=1}^{n} c_i} \cong \frac{1}{n} \tag{15}$$

n is the number of conformations and c_i is the concentration of the ith conformer. In the above derivation it is assumed that the entropy is effectively the same for all the conformers and the concentration of the conformations are approximately equal.

Based on the foregoing discussion the total entropy and heat capacity in this work were evaluated from the expressions below:

$$S^o_{(total)} = S^o_{(RRHO)} + S^o_{hr} + S^o_{mix} \tag{16}$$

$$C^o_{p(total))} = C^o_{p(RRHO)} + C^o_{p(hr)} \tag{17}$$

Computational Procedure

Calculation of the RRHO Properties

The RRHO evaluation of the thermochemical properties for the molecules and the free radicals was done using the density functional theory (DFT). The species electronic structures were optimized at the B3LYP [32-33] /6-31G** [34-35] level of theory/basis set combination. This was followed by frequency calculations based on the optimized geometries.

Most of these computations were performed using the Gaussian 09 [23a] software available in our laboratory as a central facility. However, as discussed in the Results section, similar calculations were performed using another version of the same software, namely, Gaussian 03 W [23b] in some cases. The latter version being a cheaper, portable one which could be easily run on an individual desktop/workstation is an attractive option for scientists and engineers not having access to bigger Linux based central facilities running the latest Gaussian versions. The idea behind this split exercise was to examine if acceptable results could be obtained irrespective of the software version.

The frequencies have been modified by a scaling factor of 0.9613. The geometry information including the values for the principal moments of inertia and the frequency data were used to calculate the entropy and the heat capacity under the RRHO assumption according to equations (Equations 1-10) summarized in the previous section, with appropriate manual assignment of the external symmetry numbers as required.

CALCULATION OF THE HINDERED ROTATION CONTRIBUTION TO PROPERTIES

As discussed earlier the selection of the internal rotation frequencies is a prerequisite for calculating the hindered rotation correction. In the present work, we have made the choices by systematically visualizing a range of low vibration frequencies through the graphical visualization software Gaussview [41]. We verified and confirmed our selection after carefully examining the internal rotational modes identified by the Gaussian 09 or Gaussian 03 W [23a,b] thermochemistry run using the 'freq = hinderedrotor' keyword, though we did not always go by the output of the software. Only those rotational modes that unambiguously show rotation around the bond in question were selected for the hindered rotation correction treatment.

The calculation of the "free-rotor" partition function, Q_f, requires the reduced moment of inertia (I_r). This parameter is usually calculated as shown by Pitzer and coworkers [12-13] and in the most general case of multiple rotors of any type [14,24] from the kinetic energy matrix for rotation of the molecule, converting the same into a matrix of the internal rotations. In this work we have used the results from the same Gaussian [23a.b] run, mentioned above, as the software during its hindered rotation analysis does these calculations and reports the reduced moment of inertia values for different internal rotors.

Correct symmetry number for the internal rotation is also necessary for the calculation of Q_f. In this study we have assigned the parameter value based on the standard rules [28-29], as well as by observing the potential energy surface for the particular rotation if and when required to confirm the same. We did not find the assignment in the Gaussian report always consistent and correct. From our experience in these calculations, the symmetry number (by way of influencing the "free rotor" partition function parameter to be used in the Pitzer-Gwinn tables) affects the HR corrections more significantly than the barrier height values.

The barrier height for the internal rotation of the selected hindered rotor was calculated as the difference in energy between the transition state and the corresponding energy minimum (without using the zero point energy, ZPVE, correction) after approximately locating these on the potential energy versus torsion angle diagram for each internal rotational mode. Identification of the exact transition state was accomplished by the QST2 method using Gaussian 09 [23a]. The MP2 [36-40] /6-311G** [34-35] level of theory/basis set combination was used to calculate the energy of these states. It is expected that the barrier height being calculated in this work by using a higher chemistry level than the method/basis set combination with which the structure optimization and frequency calculations were performed would make this parameter value more accurate than the corresponding value reported by the default HR analysis by the software.

The hindered rotation contribution to entropy and heat capacity were calculated from the Pitzer and Gwinn [12] tables along with the Equations (11) and (12) and added to the RRHO values. It is to be noted that while adding this contribution to the RRHO values, the corresponding contribution of the harmonic vibration at the frequency associated with the internal rotation was subtracted. The correction would be repeated over all the internal rotation modes.

Calculation of the Entropy of Mixing Correction

As mentioned in the previous section, in this study the entropy of mixing due to the presence of low energy conformers was considered as a further correction to the RRHO value of entropy (in addition to that due to the hindered rotations) as per the Equation (16). The Equation (14) was used to calculate the entropy of mixing term. To use this approximation, one has to calculate the number of conformations.

This approximation works because the exact amounts of the conformers do not matter as long as high-energy conformers are counted out. We calculated the number of conformers using a qualifying condition following Guthrie [25]. We assumed that only the conformers having an energy difference up to about 1.0 kcal/mol with respect to the lowest energy conformer (reference conformer) would contribute to the entropy of mixing term thus excluding others having energy higher than 1.0 kcal/mol in comparison to the reference conformer.

In order to determine the reference conformer we followed some general rules that are applicable to alkanes. Applicability of this basis to free radicals in a general way may be debatable. But since in this work we would be dealing with straight chain and branched alkyl radicals that admit multiple conformers we assumed that considerations as used for alkanes of similar structures would hold.

It is generally known that the *anti* conformations are lower in energy than the corresponding *gauche* conformations. Therefore, the radical conformation where all the C-C-C-C dihedral angles have a value of 180°, i.e., all are present in the *anti* conformation, has been assumed as the lowest energy conformer. In the case of the branched radicals, the conformation with the lowest number of *gauche* interactions has been assumed as the lowest energy conformer. Once the lowest energy conformer has been identified, the other conformers can be calculated with ease.

It is well known that one *gauche* interaction contributes a destabilisation energy of 0.7 kcal/mol [42] and that two *gauche* interactions will lead to a conformer that is 1.4 kcal/mol higher in energy. Therefore, conformations having more than one *gauche* interaction will be less stable than the reference conformer by more than 1.0 kcal/mol (and therefore be out of the assumed cut-off point) and hence will not be considered to contribute to the entropy of mixing term.

In other words, conformations having only one extra *gauche* interaction with respect to the reference conformer will contribute to the entropy of mixing. Taking the conformation with the maximum number of *anti* dihedrals as the reference conformer, a systematic variation of each dihedral can generate two new conformations, each being 0.7 kcal/mol higher in energy than the reference conformer. Note that only the dihedrals for the carbon-carbon backbone has been considered as dihedrals with other atoms such as hydrogen, leading to methyl tops will not lead to new, distinguishable conformations. Therefore, if there are "n" C-C-C-C dihedrals in the structure, the number of conformations to be considered will be 2n + 1.

RESULTS AND DISCUSSION

The methodology as set out in the previous sections was used to make a priori calculation of the ideal gas entropy and heat capacity at 298.15 K of a number of organic molecules and free radicals in a consistent manner.

Before presenting and discussing the results of application to a series of straight chain and branched alkyl radicals, most of them not having been studied systematically using the "3rd law" method to the best of our knowledge (nor are experimentally determined values of the thermochemical properties of these radicals readily available), it was felt that the methodology should be standardised and validated by comparison against a set of radicals and molecules which previous researchers studied by the same basic method with some variations in the implementation details and the calculated property values were published. Three sets of studies were identified for these comparisons, namely, a series of haloalkyl radicals [15-18], another set of small alkyl (both straight chain and branched) radicals and the corresponding peroxy radicals (obtained typically by addition of an oxygen molecule during oxidation processes) [19-20] and finally a series of simple hydrocarbon molecules (part of a systematic study involving a large number of various hydrocarbons [21-22]). For the first two sets of radicals only the hindered rotation (HR) corrections being pertinent were calculated, thus allowing us to compare the corrected property values obtained in this work with those presented in the literature.

On the other hand for the set of simple hydrocarbon molecules allowance was made for the presence of a mixture of conformers wherever they occur and the corresponding corrections for entropy by way of the entropy of mixing (MIX), were added to the HR corrected RRHO values and the overall values compared with those reported in the literature.

For all the radicals and the molecules considered within the above three sets, the HR corrections to the RRHO values of entropy and the heat capacity were also calculated by running the Gaussian software with the 'freq = hinderedrotor' keyword. The Gaussian report had all the necessary data reported in the tables of comparison below. However, it should be noted that the Gaussian run makes an automatic calculation of the hindered rotor corrections without any user intervention as regards the choices of the calculation parameters. The Gaussian reported corrected 'total' entropy and heat capacity do not seem to contain any correction due to the presence of conformers.

HALOALKYL RADICALS

Table 1 presents the entropy values (the RRHO component and the HR correction) for 11 α- and/or β-fluoro-/chloro-substituted ethyl radicals, as calculated within the present study and the corresponding values computed by using Gaussian 09 [23a].

Table 2 presents the heat capacity values for the same radicals. In both these tables the corrected ('total') property values obtained by both methods were compared with those computed and listed in a series of papers by Tschuikow-Roux et al [15-18].

Table 1. The entropy* with hindered rotation correction for haloalkyl radicals

Radical	Present Study			Gaussian 09 [23a]			Literature [15-18]	Diff[a]	Diff[b]
	S_{RRHO}	S_{hr}	S_{total}	S_{RRHO}	S_{hr}	S_{total}			
CH_3-C·HF	272.4	1.1	273.5	272.3	1.9	274.2	274[15]	0.5	-0.2
CH_3-C·F_2	289.5	1.1	290.6	289.3	1.6	290.9	290.3[15]	-0.3	-0.6
CH_2F-C·H_2	274.9	4.8	279.7	274.8	0	274.8	279.7[16]	0	4.9
CHF_2-C·H_2	294.3	3.2	297.5	294	0.8	294.8	297.8[16]	0.3	3
CF_3-C·H_2	309.5	-0.6	308.9	309.3	1.1	310.4	302.6[16]	-6.3	-7.8
CH_2F-C·HF	299.1	1.4	300.5	298.8	1.2	300	293.3[17]	-7.2	-6.7
CHF_2-C·HF	314.9	5.8	320.7	314.3	0.4	314.7	310.6[17]	-10.1	-4.1
CF_3-C·HF	328.7	0.4	329.1	328.4	1.9	330.3	326.2[17]	-2.9	-4.1
CH_2Cl-C·H_2	284.5	4.5	289	283.5	0	283.5	289[18]	0	5.5
CH_3-C·HCl	287.5	0.8	288.3	287.8	0.3	288.1	288.1[18]	-0.2	0
CH_3-C·Cl_2	314.9	0.1	315	315.4	1	316.4	314.6[18]	-0.4	-1.8

* Entropy values are reported in $JK^{-1}mol^{-1}$. [a] Deviation of the entropy values obtained in the present study from those reported by Tschuikow-Roux et al [15-18] and the [b]deviation of the values computed by Gaussian 09 [23a] from the same literature reported ones. The average absolute deviation of the former is 3.1 and that of the latter is 3.5 $JK^{-1}mol^{-1}$.

Table 2. The heat capacity* with hindered rotation correction for haloalkyl radicals

Radical	Present Study			Gaussian 09 [23a]			Literature [15-18]	Diff[a]	Diff[b]
	Cp_{RRHO}	Cp_{hr}	Cp_{total}	Cp_{RRHO}	Cp_{hr}	Cp_{total}			
CH_3-C·HF	60	0.5	60.5	60	0.6	60.6	58.7[15]	-1.8	-1.9
CH_3-C·F_2	67	1.2	68.2	67.1	1.3	68.4	67.2[15]	-1	-1.2
CH_2F-C·H_2	62.1	-3.6	58.6	62.5	0	62.5	57.9[16]	-0.7	-4.6
CHF_2-C·H_2	70.9	-3.8	67.1	70.9	1.1	72	66.8[16]	-0.3	-5.2
CF_3-C·H_2	81.9	-3.9	78	82	1.4	83.4	77.1[16]	-0.9	-6.3
CH_2F-C·HF	67	1.2	68.2	67	1.5	68.5	69.4[17]	1.2	0.9
CHF_2-C·HF	77.1	1.5	78.6	77.1	0.5	77.6	75.9[17]	-2.7	-1.7
CF_3-C·HF	87.8	0.7	88.5	87.8	0.4	88.2	85.6[17]	-2.9	-2.6
CH_2Cl-C·H_2	65.4	1.1	66.5	65.5	0	65.5	64.7[18]	-1.8	-0.8
CH_3-C·HCl	64.7	-0.8	63.9	64.9	0.4	65.3	61.8[18]	-2.1	-3.5
CH_3-C·Cl_2	75.6	1.1	76.7	75.7	1.4	77.1	75.8[18]	-0.9	-1.3

* Heat capacity values are reported in $JK^{-1}mol^{-1}$. [a] Deviation of the heat capacity values obtained in the present study from those reported by Tschuikow-Roux et al [15-18] and the [b]deviation of the values computed by Gaussian 09 [23a] from the same literature reported ones. The average absolute deviation of the former is 0.9 and that of the latter is 1.7 $JK^{-1}mol^{-1}$.

For all these radicals involving a single symmetric (and in some cases asymmetric) rotating top, the use by these authors of the classical Pitzer methodology and the Pizer-Gwinn tables [12] was expected to yield acceptable approximation of the HR corrections. Though they have not presented the RRHO values and the corrections separately the total values provided good benchmarks for testing the correctness of the present methodology, which for these radicals was almost identical with one used in these papers. It is therefore no surprise that the mean absolute deviation (MAD) of our heat capacity and the entropy values from the reported ones are 0.9 and 3.1 $JK^{-1}mol^{-1}$ respectively. The corresponding MAD numbers obtained using Gaussian are 1.7 and 3.5 $JK^{-1}mol^{-1}$. The individual deviation values were uniformly low for all radicals in the case of heat capacity and for entropy of most radicals except for two α- β- fluoro radicals (CH_2F-$C^{\cdot}HF$ and CHF_2-$C^{\cdot}HF$) and one β- fluoro radical (CF_3-$C^{\cdot}H_2$). Some comments on this point will follow a little later.

The Gaussian results are generally a little inferior to ours with reference to the literature report. The RRHO values are generally very close (though not exactly the same) as expected, assignment of the external rotational symmetry number for all these asymmetric radicals having presented no issues of divergence. The HR values differed in general for several reasons, one of them being the difference in the barrier height values. In this work this parameter was calculated by a different method using a higher theory level/basis set combination than the default chemistry level (same as one used for structure optimization) possibly employed in the Gaussian calculations. Also sometimes the internal symmetry numbers assigned were found to be different from our assignment and this might have caused some divergence in the computed HR correction.

Tchuikow-Roux et al [15-18] had employed a slightly different chemistry (UHF/6-31G*) for the structure optimization than used here (B3LYP/6-31G**). This might have caused marginal differences in their RRHO values (not reported in their papers) than ours. However, it would be more useful to compare our HR corrections with theirs or more particularly, the individual parameters used by them in these calculations and the corresponding ones employed in this work. This has been done in the Table 3.

To start with, the frequency associated with the internal rotation is found in this work to be generally comparable with that reported in the earlier work for most radicals (almost the same in a few cases) despite the differences in the chemistry level for the frequency calculations. That the barrier heights are quite close in both cases is only to be expected as the calculation method as well as the chemistry (MP2/6-311G**) are identical, though we have not accounted for the ZPVE in the calculation unlike Tschuikow-Roux et al [15-18]. Interestingly, these authors calculated the reduced moment of inertia by using the expressions originally suggested in the papers by Pitzer and the coworkers [12-13]. As explained in an earlier section, we had chosen to use the value of this parameter as it appeared in the report on the Gaussian analysis of the hindered rotation. There seems to be an indirect support for the validity of our approach in that the values of this parameter independently computed by the previous researchers match quite closely with those we had used.

The assignment of the internal symmetry numbers for most of these radicals with single symmetrical or asymmetrical tops appeared to be straightforward using the standard rules [28-29] or by looking at the torsional potential energy surfaces, wherever possible, generated during the calculation of the barrier heights. There appears to be unanimity in most of the assignments except occasional differences which, as expected, caused significant difference in the HR correction and hence the corrected entropy value as in the case of the CF_3-$C^{\cdot}H_2$ radical. However, interestingly, for another radical, namely, CHF_2-$C^{\cdot}HF$ despite the symmetry number being the same and the barrier height very close in both the studies, the corrected entropy calculated in this work significantly exceeded theirs.

Table 3. Parameters for the hindered rotation calculations for the haloalkyl radicals

Radical	Present Study				Literature [15-18]			
	Wave number	Symmetry number	Reduced moment of inertia	Barrier height	Wave number	Symmetry number	Reduced moment of inertia	Barrier height
	cm^{-1}		amu* $(bohr)^2$	kJ mol^{-1}	cm^{-1}		amu* $(bohr)^2$	kJ mol^{-1}
CH_3-$C^{\cdot}HF$	193	3	0.960	7.90	186	3	1.168	8.36
CH_3-$C^{\cdot}F_2$	202	3	1.178	10.07	188	3	1.206	10.42
CH_2F-$C^{\cdot}H_2$	167	2	0.629	1.10	161	2	0.623	1.03
CHF_2-$C^{\cdot}H_2$	131	2	0.679	0.93	128	2	0.676	0.53
CF_3-$C^{\cdot}H_2$	124	3	0.689	0.79	124	6	0.691	0.76
CH_2F-$C^{\cdot}HF$	63	1	2.878	16.73	104	1	2.896	10.38
CHF_2-$C^{\cdot}HF$	81	1	4.058	14.21	85	1	4.086	14.52
CF_3-$C^{\cdot}HF$	78	3	4.979	7.98	80	1	4.877	8.16
CH_2Cl-$C^{\cdot}H_2$	247	2	0.647	8.09	200	2	0.636	8.18
CH_3-$C^{\cdot}HCl$	153	3	1.020	5.69	164	3	1.018	5.62
CH_3-$C^{\cdot}Cl_2$	178	3	1.222	10.37	173	3	1.217	10.33

There is little reason for the HR corrections to be different in this case, which cannot, however, be verified as the value was not reported by the authors. Thus on the whole our calculation procedure can be said to have by and large reproduced published data on these haloalkyl radicals within the stated accuracy.

Alkyl and Alkyl Peroxy Radicals

In connection with their systematic experimental work on the kinetics and equilibria of $R^{\cdot} + O_2 \Leftrightarrow RO_2^{\cdot}$ reactions for a variety of small alkyl radicals, Slagle et al [19] also presented results of "3rd Law" computations for thermodynamic properties including entropy of some alkyl and the corresponding alkyl peroxy radicals. In a subsequent paper Knyazev and Slagle [20] had reanalyzed the earlier experimental data and apparently also recalculated the entropy for some of the radicals and presented these updated values. Though the details of their calculation procedure appear incomplete, they have used the Pizer methodology including the tables [12] for evaluation of the hindered rotor corrections for these radicals with multiple symmetric tops (as in the case of the alkyl radicals) and for radicals showing both symmetric and asymmetric tops (as for the peroxy radicals). Since we have adopted a similar approach in this work and our other applications would involve, in general, such multiple rotor situations it was of interest to compare the entropy of selected radicals calculated here with those reported by Slagle and coworkers [19-20]. Table 4 presents the entropy values (the RRHO value and the HR correction) for 4 alkyl (namely, methyl, ethyl, iso-propyl and tert-butyl) radicals calculated in this work and the corresponding values computed by using Gaussian software. The corrected ('total') property values by both methods were then compared with those reported by Slagle and coworkers [19-20]. The software version used was the Windows based Gaussian 03 W [23b] for all the radicals except for the tert-butyl peroxy radical for which due to numerical convergence problems the calculations were done with Gaussian 09 [23a]. Other trial calculations showed that although numerical values of some parameters differed the entropy results obtained with both the versions were not significantly different.

It turns out that entropies calculated here were quite comparable with the previously computed values [19-20] with a MAD of 1.8 $JK^{-1}mol^{-1}$ while the Gaussian calculations show bigger deviations for all the radicals (MAD 4.1 $JK^{-1}mol^{-1}$).

Table 4. The entropy* with hindered rotation correction for the alkyl and the alkyl peroxy radicals

Radical	Present Study			Gaussian 03 W [23b]			Literature [19,20]	Diff[a]	Diff[b]
	S_{RRHO}	S_{hr}	S_{total}	S_{RRHO}	S_{hr}	S_{total}			
$C^{\cdot}H_3$	195.4	0.0	195.4	195.5	0.0	195.5	193.7[19]	-1.7	-1.8
CH_3-$C^{\cdot}H_2$	256.3	-2.4	253.9	256.3	-1.3	255.0	247.2[20]	-6.7	-7.8
$(CH_3)_2$-$C^{\cdot}H$	293.9	-0.1	293.8	292.3	2.1	294.4	293.1[19] 289.4[20]	-0.7 -4.4	-1.3 -5.0
$(CH_3)_3$-$C^{\cdot}$	316.9	2.7	319.6	316.9	5.2	322.1	317.1[19] 314.0[20]	-2.5 -5.6	-5.0 -8.1
CH3- $OO^{\cdot}$	269.7	0.6	270.3	269.7	0.8	270.5	268.6[19]	-1.7	-1.9
CH_3-CH2-$OO^{\cdot}$	303.7	10.8	314.5	303.7	2.0	305.7	311.8[20]	-2.7	6.1
$(CH_3)_2$-CH- $OO^{\cdot}$	326.9	12.3	339.2	326.9	3.6	330.5	338.7[20]	-0.5	8.2
$(CH_3)_3$-C- $OO^{\cdot}$	354.1	4.5	358.6	354.2‡	5.4‡	359.6‡	353.9[20]	-4.7	-5.7

* Entropy values are reported in $JK^{-1}mol^{-1}$. ‡ For this radical the computations were performed with Gaussian 09 [23a]. [a] Deviation of the entropy values obtained in the present study from those reported by Slagle et al. [19] and Knyazev et al [20] and the [b]deviation of the values computed by Gaussian 03 W [23b] from the same literature reported ones. The average absolute deviation of the former is 1.8 and that of the latter is 4.1 $JK^{-1}mol^{-1}$.

Though the previous researchers used the same basic HR calculation methodology, there were some differences in details. Table 5 summarises the parameters for the HR calculations used in the present study for six radicals. The table also lists two key parameters reportedly used by Knyazev and Slagle [20] in each case. The symmetry numbers assigned in both study matched (except for the ethyl radical), which might have contributed to the closeness of the values of the HR correction, hence the corrected entropies. We had arrived at the symmetry numbers mainly based on the arguments that the methyl rotors have threefold symmetry (as periodic triple wells on the potential energy surface would show) while the C-O_2 top is an asymmetric one. This latter argument holds for the methyl, ethyl and iso-propyl peroxy radicals as the torsional angle vs the potential energy diagrams also show up the asymmetry of this top. But a similar diagram for the tert-butyl peroxy radical suggests (as also stated by Knyazev and Slagle [20]) that the symmetry number should be 3 instead of 1.

On the other hand the barrier heights for some rotors (e.g., C-O_2) in some radicals (e.g., iso-propyl peroxy and tert-butyl peroxy radicals) differed significantly between the two studies.

Table 5. Parameters for the hindered rotation calculations for the alkyl and the alkyl peroxy radicals

Radical	Present Study				Literature [20]	
	Wave number	Symmetry number	Reduced moment of inertia	Barrier height	Symmetry number	Barrier height
	cm^{-1}		amu* $(bohr)^2$	kJ mol^{-1}		kJ mol^{-1}
CH_3-$C^{\cdot}H_2$	126	3	0.445	1.69	6	0.0
$(CH_3)_2$-$C^{\cdot}H$	114	3	1.218	3.42	3	3.05
	129	3	0.803	3.50	3	
$(CH_3)_3$-$C^{\cdot}$	133	3	1.157	4.97	3	6.38
	133	3	1.157	4.97	3	
	139	3	1.248	4.58	3	
CH_3-CH2-$OO^{\cdot}$	71	1	3.011	2.88	1	6.0
	224	3	1.048	10.43	3	11.3
$(CH_3)_2$-CH- $OO^{\cdot}$	119	1	4.491	12.52	1	7.17
	206	3	1.226	10.28	3	15.5
	243	3	1.21	13.97	3	
$(CH_3)_3$-C- $OO^{\cdot}$	123	3	4.304	12.84	3	10.07
	181	3	1.278	8.18	3	16.03
	234	3	1.203	12.74	3	
	246	3	1.230	14.48	3	

More importantly, Knyazev and Slagle [20] had used the identical barrier height for all the methyl tops and taken the multiple of the HR contribution of just one rotor (i.e., twice or thrice), whereas we had calculated the barrier for each identified rotor, calculated the separate HR contribution for each of them and taken a sum to find the overall contribution. Overall, the methodology (especially the identification and calculation of the HR related parameters) used in the present work seemed to be reliable in predicting the entropy of these moderately complex free radicals, which have great practical relevance. Though not presented, because comparable literature reported values were not available, heat capacities were also predicted (just like in the case of haloalkyl radicals) by the same set of calculations as above.

ALKANES

De Tar [21-22]'s implementation of the basic "3rd Law" method (that other researchers had used) for a priori estimation of entropy and heat capacity of a large set of hydrocarbon molecules had the following distinguishing features. He (a) used, unlike other previous workers, a relatively high theory level/basis set combination to calculate the RRHO properties with very good accuracy; (b) like others, employed the Pizer and Gwinn [12] methodology to approximate the HR corrections (to be added to the RRHO values) for the internal rotation of the multiple methyl groups present in the molecules (tried to compute accurate barrier heights again with high theory level/basis set combinations); and finally (c) devised a procedure to compute a correction to both the properties to account for the realistic situation of a compound existing as a mixture of stable conformers rather than only as a minimum energy configuration. Molecules considered being hydrocarbons and having access to at least some experimentally determined property data the author could go to great lengths in implementing more accurate computational techniques. The procedure to correct properties due to mixture of conformers is also quite elaborate which allowed for variation of the properties among the conformers. The claimed accuracy vis-à-vis available experimental data and the precision of these theoretical calculations were impressive.

Since we were interested in standardising a version of the "3rd Law" method that would be more easily adaptable to a range of free radicals we hoped to validate our methodology (which deployed a more modest computational effort than above) by applying the same to known examples of radicals and molecules for which similar calculations yielded reliable results. So far we showed evidence of successful application to several radicals with both symmetrical and asymmetrical internal rotations, presence of multiple similar and dissimilar rotors within the same radical. In none of the previous examples we dealt with, the presence of multiple conformers was relevant. In what follows we present some results (with reference to the data provided in De Tar [22]) showing how our methodology fared for a small set of selected hydrocarbons where both our adaptation of the Pitzer methodology for the HR calculations and the use of the simplified approach due to Guthrie [25] to calculate the entropy of mixing correction due to conformers when present would be tested.

Table 6 presents the entropy values (the RRHO component, the HR and the entropy of mixing corrections) for 5 simple alkanes, calculated with the present methodology and the corresponding values computed by using Gaussian 03 W [23b]. Table 7 presents the heat capacity values for the same molecules. The latter property was calculated without any correction for the mixture of conformers. In these tables the corrected ('total') property values by both methods were compared with those presented by De Tar [22]. To make closer comparison of the effectiveness and accuracy of the HR correction in each case, in the Table 8 the usual hindered rotation calculation parameters used in our study as well as the resulting HR corrections for entropy were compared with those reported in De Tar [22].

From the Table 6 and 7 it appears that for the alkanes considered the deviations of the entropy and heat capacity values calculated in this work from those in De Tar [22] in most cases (entropy of butane is an exception) are not significant. As expected despite the difference in the chemistry level the RRHO values differ only very marginally. On the other hand, Table 8 shows that the HR corrections here always exceeded the previously calculated values (most dramatically for n-butane).

Table 6. The entropy* with hindered rotation and entropy of mixing corrections for alkanes

Alkane	Present study					Gaussian 03 W [23b]			De Tar [22]	Diff [a]	Diff [b]
	$n^{\dagger}$	S_{RRHO}	S_{hr}	S_{mix}	S_{total}	S_{RRHO}	S_{hr}	S_{total}			
Ethane	1	228.2	1.4	0	229.6	228.3	1.3	229.6	229	-0.6	-0.6
Propane	1	269.8	2.4	0	272.2	269.8	2.3	272.1	271	-1.2	-1.1
iso-Butane	1	294.8	3.3	0	298.1	294.8	3.0	297.9	296.3	-1.8	-1.6
n-Butane	3	301.9	9	9.1	320	301.9	12.8	314.7	311.9	-8.1	-2.9
n-Pentane	5	334.6	2.5	13.4	350.5	334.6	10.6	345.2	351.4	0.9	6.2

* Entropy values are reported in $JK^{-1}mol^{-1}$. † Number of conformations.

[a] Deviation of the entropy values obtained in the present study from those reported by De Tar [22] and the [b]deviation of the values computed by Gaussian 03 W [23b] from the same literature reported ones.

Table 7. The heat capacity* with hindered rotation correction for alkanes

Alkane	Present Study			Gaussian 03 W [23b]			De Tar [22]	Diff [a]	Diff [b]
	Cp_{RRHO}	Cp_{hr}	Cp_{total}	Cp_{RRHO}	Cp_{hr}	Cp_{total}			
Ethane	51.7	1.4	53.1	51.7	1.8	53.5	52.9	-0.2	-0.6
Propane	72.2	2.8	75.0	72.2	3.4	75.6	74.7	-0.3	-0.9
iso-Butane	95.5	2.6	98.1	95.6	4.3	99.9	98.9	0.8	-1.0
n-Butane	93.0	6.0	99.0	93.0	4.6	97.6	97.8	-1.2	0.2
n-Pentane	114.3	2.8	117.1	114.3	6.6	120.9	119.3	2.2	-1.6

* Heat capacity values are reported in $JK^{-1}mol^{-1}$. [a] Deviation of the heat capacity values obtained in the present study from those reported by De Tar [22] and the [b]deviation of the values computed by Gaussian 03 W [23b] from the same literature reported ones.

As stated by De Tar [22] his HR correction was found by calculating the same for a single methyl top and multiplying resulting value by the number of tops in the molecule assuming that the contribution is same for all the rotors. Table 8 shows that our procedure calculated barrier height and the reduced moment of inertia for each identifiable rotational motion of different methyl tops, though some of these parameters were numerically comparable in both studies. The Gaussian calculated HR values for both properties of all the alkanes (except for n-pentane) were fairly close to ones calculated by our procedure.

The entropy of mixing calculated in the present study (for n-butane and n-pentane) are not directly comparable with the corrections calculated in the earlier work, as ours provide only one part of the correction (due to presence of a mixture of conformers with the same entropy values), that too using a simple approximate formula.

However, on comparing the mixing part of this correction as reported in De Tar [22], for n-butane this is 7.9 as against 9.1 in this work and for n-pentane the corresponding numbers are 11.2 and 13.4. This indicates that the approximation used in this work for the entropy of mixing correction may produce results not too much off the mark.

Table 8. Parameters for calculation of the hindered rotation correction for the entropy of alkanes

Molecule	Present Study				Literature [22]			
	Wave number	Reduced moment of inertia	Barrier height	S_{hr}	Wave number	Reduced moment of inertia	Barrier height	S_{hr}
	cm^{-1}	amu* $(bohr)^2$	$kJ\ mol^{-1}$	$JK^{-1}mol^{-1}$	cm^{-1}	amu* $(bohr)^2$	$kJ\ mol^{-1}$	$JK^{-1}mol^{-1}$
Ethane	313	0.624	2.67	1.4	295	0.614	2.96	0.8
Propane	223 274	1.212 0.931	2.68 3.22	2.4	207 262	1.073	3.28	1.0
iso-Butane	216 264 264	1.243 1.170 1.170	2.55 3.69 3.69	3.3	200 249 249	1.185	3.60	1.2
n-Butane	220 256	1.175 1.075	2.69 3.54	9.0	217 250	1.063	3.11	1.7
n-Pentane	242 250	1.026 1.052	2.74 2.97	2.5	230 240	1.063	3.11	2.1

C_4-C_6 Alkyl Radicals

In the remainder of this chapter, we consider a series of 24 larger alkyl radicals with 4 to 6 carbon atoms, many of them with substantial branched structures. In a classic paper published many years ago, Cohen [43] had presented the ideal gas entropy at 298 K and the heat capacity data between 298-1500 K having calculated the properties by an approximate estimation procedure, named "difference method", originally devised by O'Neal and Bension [6]. We had made a priori calculation of the properties for these radicals using the methodology standardised in this chapter and compared our results with the data presented by Cohen [43]. For each radical the properties were also calculated by using Gaussian 09 [23a].

COMPARISON WITH THE PUBLISHED DATA

Table 9 compares for each of these 24 radicals, the entropy, calculated in the present study as well as using Gaussian, against the values listed in Cohen [43]. Table 10 presents a similar comparison for the heat capacity. Unlike Cohen's study, however, both the present study and Gaussian computations provided a break up of a property value in terms of the components that go to make it – the RRHO value and the HR correction. Additionally, in the case of entropy, for a radical existing as a mixture of conformers, the entropy of mixing correction also formed a part of the total property value in the present study, but not in the Gaussian calculations. Table 9 also contains a column (the first one) listing the pertinent number of conformers obtained by the enumeration procedure detailed in an earlier section of this chapter. This column reports a value of unity for about 9 radicals like 1,2-dimethyl butyl or 1,1,2-trimethyl propyl, which exist as a single conformer.

The last two columns in both these tables list the deviation of the corrected property values calculated here and by using Gaussian from those reported by Cohen [43]. The MAD values for entropy are 3.0 and 5.4 $JK^{-1}mol^{-1}$ respectively and those for heat capacity are 3.2 and 4.5 $JK^{-1}mol^{-1}$. These literature values are not experimentally determined ones but obtained by an approximate method and using them as comparison benchmarks and to comment on the predictive ability of the present methodology for these radicals per se may be debatable. However, for the limited purpose of critically examining its strength and the limitations in a general sense, these comparisons would be worthwhile.

Table 9. The entropy* with hindered rotation and entropy of mixing correction for C_4-C_6 alkyl radicals

Radical	Present Study				Gaussian 09 [23a]			Cohen	Diff[a]	Diff[b]
	n_{conf}	S_{RRHO}	S_{hr}	S_{mix}	S_{total}	S_{RRHO}	S_{hr}	S_{total}	[43]		
1-Butyl	3	321.7	9.5	9.1	340.3	321.7	12.1	333.8	331.4	-8.9	-2.4
2-Butyl	3	327.4	0.2	9.1	336.7	327.4	2.9	330.3	331.1	-5.6	0.8
1-Pentyl	5	353.5	11.8	13.4	378.7	353.6	13.5	367.1	370.2	-8.5	3.1
2-Pentyl	5	359.6	5.0	13.4	378.0	359.6	23.5	383.1	369.9	-8.1	-13.2
1-Hexyl	7	386.1	12.7	16.2	415.0	386.1	6.8	392.9	409.7	-5.3	16.8
2-Hexyl	7	392.6	5.1	16.2	413.9	392.6	7.1	399.7	409.4	-4.5	9.6
1-Ethyl propyl	3	361.7	2.1	9.1	372.9	361.8	14.5	376.3	364.7	-8.2	-11.6
2-Methyl butyl	3	349.9	10.4	9.1	369.4	350.0	10.1	360.1	364.8	-4.6	4.7
1,2-Dimethyl Propyl	1	358.1	0.2	0.0	358.3	358.1	2.2	360.3	358.8	0.5	-1.5
3-Methyl Butyl	1	347.5	15.3	0.0	362.8	347.5	10.2	357.7	359.1	-3.7	1.4
1-Ethyl Butyl	7	393.7	1.5	16.2	411.4	393.8	16.2	410.0	409.9	-1.5	-0.1
2-Methyl Pentyl	5	380.4	10.4	13.4	404.2	380.4	14.2	394.6	401.8	-2.4	7.2
1,1-Dimethyl butyl	1	405.6	-1.5	0.0	404.1	405.6	8.8	414.4	402.2	-1.9	-12.2
1,3-Dimethyl butyl	3	387.7	1.5	9.1	398.3	387.8	11.6	399.4	395.8	-2.5	-3.6
4-Methyl pentyl	3	380.8	10.7	9.1	400.6	380.8	5.3	386.1	396.0	-4.6	9.9
3-Methyl Pentyl	5	378.0	11.6	13.4	403.0	378.0	10.6	388.6	401.0	-2.0	12.4
1,2-Dimethyl butyl	1	387.8	6.5	0.0	394.3	387.8	5.4	393.2	395.0	0.7	1.8
1-Methyl, 1-ethyl propyl	3	405.0	-1.6	9.1	412.6	405.1	1.3	406.4	401.5	-11.1	-4.9
2-ethyl butyl	3	381.6	10.2	9.1	400.9	381.6	11.4	393.0	395.3	-5.6	2.3
2,3-dimethyl butyl	1	373.7	16.6	0.0	390.3	373.8	11.1	384.9	387.0	-3.3	2.1
1,1,2-trimethyl propyl	1	389.1	1.3	0.0	390.4	389.1	5.9	395.0	393.2	2.8	-1.8
2,2 dimethyl butyl	1	374.5	11.4	0.0	385.9	374.5	10.4	384.9	382.9	-3.0	-2.0
1,2,2-trimethyl propyl	1	382.9	0.3	0.0	383.2	382.9	10.8	393.7	388.4	5.2	-5.3
3,3-dimethyl butyl	1	371.2	11.5	0.0	382.7	371.3	11.0	382.3	382.9	0.2	0.6

* Entropy values are reported in $JK^{-1}mol^{-1}$. [a] Deviation of the entropy values obtained in the present study from those reported by Cohen [43] and the [b] deviation of the values computed by Gaussian 09 [23a] from the same literature reported ones. The average absolute deviation of the former is 3.0 and that of the latter is 5.4 $JK^{-1}mol^{-1}$.

Looking at the deviation for individual radicals, for the entropy, the newly calculated values (our method) are generally higher than those from the previous study (though in about a fourth of the cases the values are underpredicted. In the case of the Gaussian calculations the deviations are more evenly distributed. Interestingly, for those 9 radicals considered as single conformers, deviations of our values from Cohen's [43] are evenly distributed and the MAD for these 9 radicals is 2.4 $JK^{-1}mol^{-1}$. As a matter of fact in Cohen's method based as it was on the differences in the molecular characteristics that may contribute to entropy between the radical and the parent molecule the issue of a conformer mixture did not seem to have been considered. Thus Cohen's data [43] basically refer to entropy values corrected as they should be due to hindered internal rotations only.

Table 10. The heat capacity* with hindered rotation correction for C_4-C_6 of alkyl radicals

Radical	Present Study			Gaussian 09 [23a]			Cohen [43]	Diff[a]	Diff[b]
	Cp_{RRHO}	Cp_{hr}	Cp_{total}	Cp_{RRHO}	Cp_{hr}	Cp_{total}			
1-Butyl	97.3	-3.4	93.9	97.3	-3.6	93.7	95.5	1.6	1.8
2-Butyl	96.4	-1.8	94.6	96.4	-4.1	92.3	97.2	2.6	4.9
1-Pentyl	118.3	-3.4	114.9	118.3	-2.1	116.2	118.1	3.2	1.9
2-Pentyl	117.3	-3.4	113.9	117.3	-3.3	114.0	119.7	5.8	5.7
1-Hexyl	139.3	-3.8	135.5	139.3	-0.4	138.9	140.6	5.1	1.7
2-Hexyl	138.4	-3.3	135.1	138.5	-2.2	136.3	142.3	7.2	6.0
1-Ethyl propyl	116.5	2.8	119.3	116.5	-3.0	113.5	119.4	0.1	5.9
2-Methyl butyl	120.6	-2.7	117.9	120.6	0.1	120.7	116.9	-1.0	-3.8
1,2-Dimethyl Propyl	119.8	-8.6	111.2	119.8	-3.7	116.1	118.6	7.4	2.5
3-Methyl Butyl	120.6	-7.2	113.4	120.6	1.2	121.8	116.9	3.5	-4.9
1-Ethyl Butyl	137.6	0.4	138.0	137.6	-1.3	136.3	141.9	3.9	5.6
2-Methyl Pentyl	141.4	-3.9	137.5	141.4	2.0	143.4	140.2	2.7	-3.2
1,1-Dimethyl butyl	139.6	-2.1	137.5	139.6	-6.5	133.1	139.6	2.1	6.5
1,3-Dimethyl butyl	140.6	-5.0	135.6	140.6	-3.9	136.7	141.9	6.3	5.2
4-Methyl pentyl	141.7	-6.5	135.2	141.7	0.6	142.3	140.2	5.2	-2.1
3-Methyl Pentyl	141.4	-1.2	140.2	141.4	2.6	144.0	138.2	-2.0	-5.8
1,2-Dimethyl butyl	141.0	-4.0	137.0	141.0	-1.3	139.7	139.8	2.8	0.1
1-Methyl, 1-ethyl propyl	139.1	-1.3	137.8	139.1	-6.1	133.0	137.5	-0.3	4.5
2-ethyl butyl	142.0	-0.5	141.5	142.0	1.7	143.7	138.2	-3.3	-5.5
2,3-dimethyl butyl	143.9	-8.7	135.2	143.9	2.8	146.7	137.4	2.2	-9.3
1,1,2-trimethyl propyl	141.4	-6.7	134.7	141.4	-3.9	137.5	136.8	2.1	-0.7
2,2 dimethyl butyl	145.9	-0.7	145.2	145.9	1.5	147.4	139.5	-5.7	-7.9
1,2,2-trimethyl propyl	145.6	-1.0	144.6	145.6	0.3	145.9	141.2	-3.4	-4.7
3,3-dimethyl butyl	145.5	4.9	150.4	145.5	0.5	146.0	139.5	-10.9	-6.5

* Heat capacity values are reported in $JK^{-1}mol^{-1}$. [a]Deviation of the heat capacity values obtained in the present study from those reported by Cohen [43] and the [b]deviation of the values computed by Gaussian 09 [23a] from the same literature reported ones. The average absolute deviation of the former is 3.2 and that of the latter is 4.5 $JK^{-1}mol^{-1}$.

The point being made is that for radicals where entropy of mixing correction is not an issue, our calculations predict entropy values quite close to those by Cohen's method [43]. However, it may not be fair to generalise this claim. If we consider only HR correction for all the 24 radicals, our values are generally lower than Cohen's with the overall MAD of about 4. In the case of heat capacity also, for which conformer mixture has not been considered in any case the majority of the values calculated here (Table 10) are lower than Cohen's.

Be that as it may, if we disregard the conformer mixtures, the HR corrected property values calculated here can be said to be closely comparable with available published values (obtained by an approximate method that may be difficult to generalise), though slightly underpredicting them. In addition our methodology allows addition of a correction due to presence of conformers to the entropy values. Based on our experience with other radicals and molecules this added correction should make our values more accurate.

HR CALCULATION PARAMETERS

While comparing the results of the present study with those obtained with Gaussian 09 [23a] we notice the significant differences in the values of HR corrections (RRHO values being practically the same). This is amply clear from perusal of the S_{hr} column in the Table 9 and the Cp_{hr} column in the Table 10. As mentioned in an earlier section, though we had used in our calculations one of the key parameters, namely, the reduced moment of inertia, for each group identified as a hindered rotor from the Gaussian HR analysis report, our reference to Gaussian calculation in this regard, which is an in-built automatic procedure within the software not quite amenable to the user intervention, ended there. Though the software reports symmetry number and the barrier height we have not used these values. We independently assigned symmetry numbers as stated before. The barrier heights were calculated by a separate procedure wherein we determined the exact rotational transition states employing the QST2 approach [44-45] using Gaussian 09 [23a] with higher theory level/basis set combination than used for geometry optimization. For each identified rotating group there may be several transition states along the potential energy surface. The procedure used will be illustrated below for the case of 1-butyl radical.

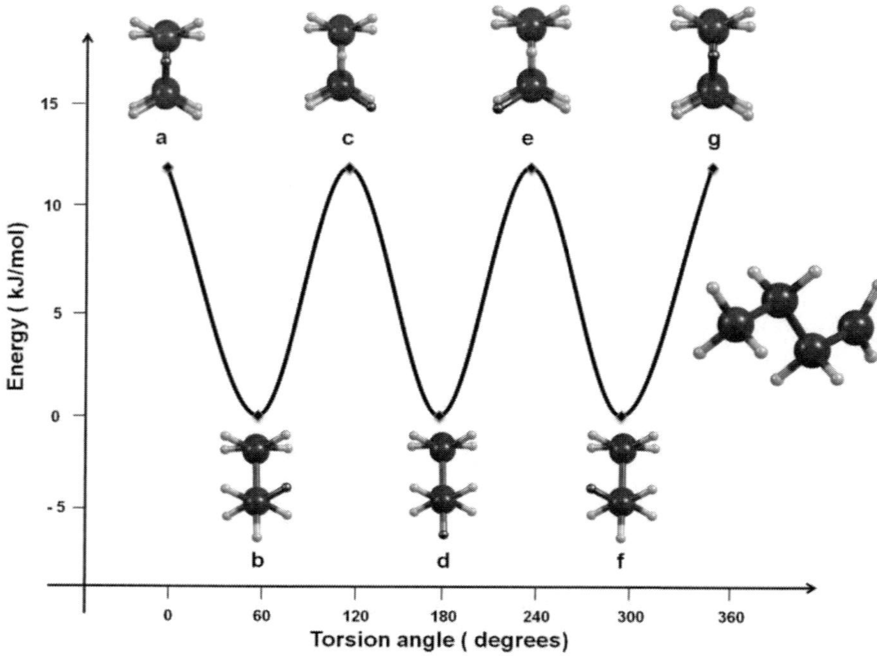

Figure 1. The potential energy diagram for the rotation of the terminal CH_3 rotor in 1-butyl radical.

In the case of 1-butyl radical, there are two hindered rotations to be considered, namely, that of the terminal CH_3 group and the $^{\cdot}CH_2$ group containing the radical center. The potential energy diagrams for the rotation of these two rotors are shown in the Figures 1 and 2 respectively. Figure 1 shows the rotation of the CH_3 rotor with a completely symmetrical periodic potential. The intermediate states have been calculated at 60° rotations, one of the hydrogens being colored black in order to facilitate the visualization of the rotation. The three minima b, d and f are not only energetically but also geometrically equivalent. The same is true for the transition states a, c and e. Hence in order to determine the barrier height in this case the determination of the exact transition state is necessary for just one among these three states. The difference in energy between the transition state and the corresponding minimum is the barrier height for this rotation.

The potential energy diagram for the $^{\cdot}CH_2$ rotor (Figure 2) has been drawn with the states at 30° rotations, one of the hydrogens colored black to facilitate the visualization. This diagram shows considerable asymmetry in the potential function with a number of local minima such as d and j or f and h at slightly higher energy levels than that corresponding to the staggered structure b and multiple transition states corresponding to these minima.

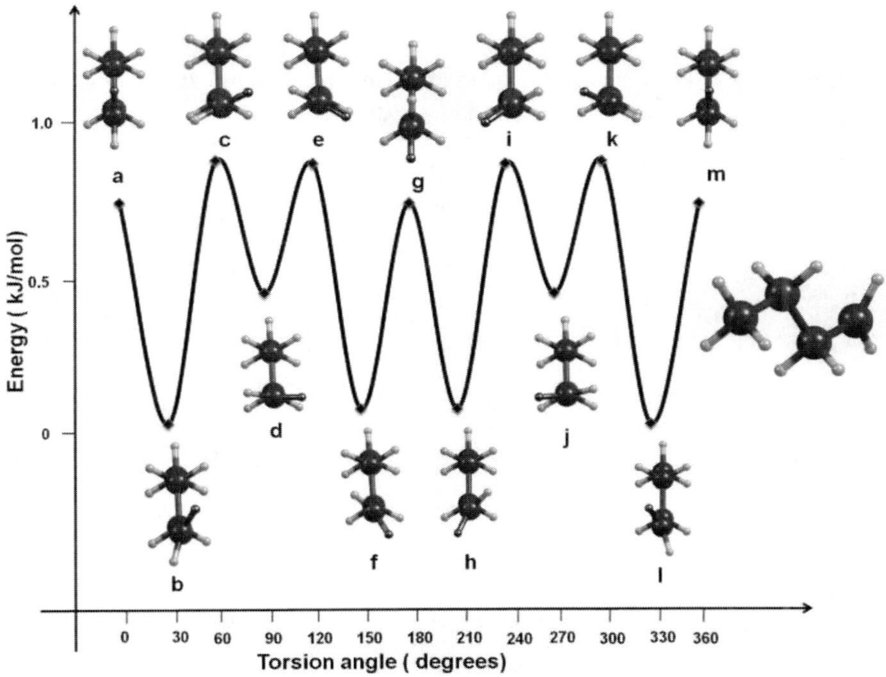

Figure 2. The potential energy diagram for the rotation of the $CH_2^{\bullet}$ rotor in 1-butyl radical.

All these states were calculated with QST2 method and the corresponding barrier height evaluated in each case. As a rule we have accepted the lowest barrier height as that indicated the minimum energy required by the hindered rotor in order to complete the rotation.

The potential energy diagrams (not presented, very similar to the Figure 1) in the case of 2-butyl radical, with two symmetric terminal methyl rotors, however show that for both these rotors considering a single transition state and hence a single barrier height would suffice, though the values were different for each rotor. The same procedure was followed systematically in the case of more highly branched alkyl radicals with many more rotating groups.

Table 11 summarises all the required parameters for the HR calculations for the alkyl radicals considered in this study. For a given radical each of the low vibrational frequencies associated with the hindered rotation of specific CH_3 or CH_2 tops were identified by careful visualization using GaussView software [41].

Table 11. Parameters for the hindered rotation calculations for C_4-C_6 alkyl radicals

Radical	Wave number	Symmetry number	Reduced moment of inertia	Barrier height
	cm^{-1}		amu* $(bohr)^2$	kJ mol^{-1}
1-Butyl	102	1	1.070	068
	242	3	1.043	12.90
2-Butyl	113	3	1.007	2.57
	236	3	1.011	13.22
1-Pentyl	102	1	1.783	0.66
	249	3	1.038	12.92
2-Pentyl	99	3	3.279	1.03
	249	3	1.044	12.95
1-Hexyl	98	1	2.430	0.64
	247	3	1.037	13.10
2-Hexyl	95	3	3.780	0.99
	246	3	1.037	12.89
1-Ethyl propyl	211	3	1.153	11.46
	244	3	1.065	11.46
2-Methyl butyl	122	1	0.7	1.80
	217	3	1.2	11.53
	250	3	1.160	13.64
1,2-Dimethyl Propyl	101	3	1.116	2.32
	231	3	1.216	14.70
	241	3	1.206	14.70
3-Methyl Butyl	120	1	3.286	2.3
	226	3	1.240	14.32
	239	3	1.184	14.39
1-Ethyl Butyl	241	3	1.052	13.36
	244	3	1.056	12.79
2-Methyl Pentyl	123	1	0.696	1.80
	240	3	1.188	13.63
	257	3	1.082	13.11
1,1-Dimethyl butyl	103	3	1.226	4.88
	103	3	1.226	4.88
	237	3	1.058	12.74

Radical	Wave number	Symmetry number	Reduced moment of inertia	Barrier height
	cm^{-1}		amu* $(bohr)^2$	kJ mol^{-1}
1,3-Dimethyl butyl	83	3	1.995	2.66
	246	3	1.231	14.36
	248	3	1.190	15.19
4-Methyl pentyl	130	1	0.661	0.74
	231	3	1.260	13.50
	256	3	1.196	14.52
3-Methyl Pentyl	127	1	0.668	2.10
	226	3	1.220	11.37
	269	3	1.132	12.96
1,2-Dimethyl butyl	91	3	6.537	2.25
	213	3	1.211	12.06
	222	3	1.171	14.31
1-Methyl, 1-ethyl propyl	70	3	1.271	2.49
	217	3	1.230	12.42
	237	3	1.119	12.84
2-ethyl butyl	124	1	0.695	2.20
	209	3	1.202	10.53
	223	3	1.142	12.26
2,3-dimethyl butyl	85	1	7.816	0.82
	217	3	1.219	14.75
	219	3	1.246	13.95
	242	3	1.206	13.53
1,1,2-trimethyl propyl	109	3	1.256	4.91
	119	3	1.219	3.96
	220	3	1.224	13.76
	260	3	1.224	14.52
2,2 dimethyl butyl	115	1	0.697	1.04
	214	3	1.213	11.0
	262	3	1.225	15.74
	279	3	1.193	15.74
1,2,2-trimethyl propyl	101	3	1.097	1.48
	218	3	1.246	16.38
	234	3	1.225	15.84
	284	3	1.225	15.84
3,3-dimethyl butyl	125	1	0.786	2.48
	230	3	1.263	16.67
	277	3	1.186	17.1
	280	3	1.226	17.1

For multiple occurrences of the methyl tops attempt was made, as far as possible, to associate each with a specific frequency, though sometimes the rotations appeared somewhat coupled. This apparent independence of the rotors was one of the assumptions made in doing the HR calculations in this work. Symmetry number was assigned for each top in a radical and these numbers generally agree with those reported by Gaussian though there were occasional mismatch. The barrier heights calculated in this work and listed in Table 11 were generally quite different from those appearing in the Gaussian report.

CONCLUSION

The study reported in this chapter demonstrated that it is possible to put together a general procedure based on consistent theoretical premises for making a priori estimates of ideal gas entropy and heat capacity of organic molecules and free radicals with acceptable accuracies. Obtaining the RRHO component of the properties has become routine with the availability of many modern quantum chemistry software like Gaussian. However, in this work we had used the standard statistical thermodynamic relationships to evaluate the RRHO part of the properties based on the structural and the vibrational frequency information for the molecular geometry optimized by Gaussian.

A software like Gaussian does also allow a user an automatic way to calculate the corrections of the property values due to hindered rotations at some low vibrational frequencies. Our experience suggests, however, that caution should be exercised before accepting the results of calculations by the software in this regard, especially as there seems to be no scope within the software for user intervention with regard to any of the parameters used in the calculation. The HR correction, in this work, was independently obtained by using the classical Pitzer method [12] and generalising its usage to a host of radicals (in different classes) containing multiple rotating tops, either symmetric or asymmetric, with single or repeat occurrences of a given top within the same radical. For a number of radicals (e.g., alkyl, haloalkyl and peroxy radicals) and a series of small alkanes (C_1-C_5) studied by previous researchers using basically a similar theoretical approach though with variations, especially, in the HR correction calculation procedure, the corrected property values, published earlier, were found to be closer to ones calculated with the present methodology than by using Gaussian software's HR correction facility. The systematic identification of the low vibration frequencies associated with each internal rotation, assignment of the corresponding symmetry number, locating the various possible transition states on the rotational potential energy surface and the calculation of an accurate barrier height allow in this new methodology a clear assessment of the influence of each of these parameters on the HR correction. Wherever published data on the corrections were available these were compared with those calculated here. In cases where they are not comparable, this could be traced to one or more parameters for HR correction and their assignment or calculation. The predictive capability of the new methodology was tested in the case of a series of 24 straight-chain and branched C_4-C_6 alkyl radicals for which the property prediction by "3rd Law" method has not been in evidence in the literature. However, the published property values estimated by an approximate method were available. The values of the HR-corrected properties predicted by the present methodology were found quite close to the published data (closer than obtained by Gaussian), though generally slightly underpredicting them. For molecules and radicals that may exist as multiple low-energy stable conformers, an additional correction due to the presence of a mixture of conformers should be added to the RRHO property. In this work this had been used in an approximate manner for entropy (i.e, entropy of mixing) of alkanes and C_4-C_6 alkyl radicals wherever relevant. It turns out that for a few alkanes considered the approximate entropy of mixing correction was reasonably close to the values obtained by a much more rigorous method used in a well regarded previous work. The correction calculated in the case of C_4-C_6 alkyl radicals modified our HR-corrected property values and it is expected that this would make the overall values more accurate than the published approximate data.

ACKNOWLEDGMENT

The authors are grateful to the Center of Excellence in Scientific Computing (COESC), National Chemical Laboratory (NCL), Pune, for providing the computational facilities used in this work. KV and YD also acknowledge the financial assistance provided under the project CSC0129.

REFERENCES

[1] W. Tsang, *Data Science Journal* 3, 1 (2004).
[2] R. P. O'Connor, L. D. Schimdt, O. Detschmann, *AIChE J.* 48, 1241 (2002).

[3] R. Atkinson, D. L. Baulch, R. A. Cox, R. F. Hampson, J. A. Kerr, M. J. Rossi, J. J. Troe, Evaluated kinetic, photochemical and heterogeneous data for atmospheric chemistry (suppl. V), IUPAC Subcommittee on Gas Kinetic Data Evaluation for Atmospheric Chemistry. *J. Phys. Chem. Ref. Data* 26, 521 (1997).

[4] J. G. Barry Halliwell, *Free Radicals in Biology and Medicine*, OUP, Oxford (2007).

[5] G. N. Lewis, M. Randall, K. S. Pitzer, L. Brewer, *Thermodynamics*; 2nd ed., McGraw Hill Book Company, New York (1961).

[6] H. E. O' Neal, S. W. Benson, *Thermochemistry of free radicals* in J. K. Kochi, Ed. "Free radicals", Wiley, New York (1973) pp319-327.

[7] G. Katzer and A. F. Sax, *J. Phys. Chem. A* 106, 7204 (2002).

[8] D. A. Clabo Jr, W. D. Allen, R. B. Remington, Y. Yamaguchi, H. F. Schaefer Iii, *Chem. Phys.* 123, 187 (1988).

[9] L. Catoire, M. T. Swihart, S. Gail, P. Dagaut, *Int. J. Chem. Kinet.* 35, 453 (2003).

[10] G. Katzer and A. F. Sax, *J. Comput. Chem.* 26, 1438 (2005).

[11] J. Pfaendtner, X. Yu, L. Broadbelt, *Theor. Chem. Acc.* 118, 881 (2007).

[12] K. S. Pitzer and W. D. Gwinn, *J. Chem. Phys.* 10, 428 (1942).

[13] K. S. Pitzer, *J. Chem. Phys.* 14, 239 (1946).

[14] J. E. Kilpatrick and K. S. Pitzer, *J. Chem. Phys.* 17, 1064 (1949).

[15] Y. Chen, A. Rauk, E. Tschuikow-Roux, *J. Chem. Phys.* 93, 1187 (1990).

[16] Y. Chen, A. Rauk, E. Tschuikow-Roux, *J. Chem. Phys.* 93, 6620 (1990).

[17] Y. Chen, A. Rauk, E. Tschuikow-Roux, *J. Chem. Phys.* 94, 7299 (1991).

[18] Y. Chen and E. Tschuikow-Roux, *J. Phys. Chem.* 96, 7266 (1992).

[19] I. R. Slagle, E. Ratajczak, D. Gutman,. *J. Phys. Chem.* 90, 402 (1986). [See also the previous papers in this series for other alkyl and peroxy radicals]

[20] V. D. Knyazev and I. R. Slagle, *J. Phys. Chem. A* 102, 1770 (1998).

[21] D. F. DeTar, *J. Phys. Chem. A* 102, 5128 (1998).

[22] D. F. DeTar, *J. Phys. Chem. A* 111, 4464 (2007).

[23] [a] M. J. Frisch, G. W. Trucks, H. W. Schlegel, G. E. Scuseria, M. A. Robb, J. R. Cheeseman, G. Scalmani, V. Barone, B. Mennucci, G. A. Petersson, H. Nakatsuji, M. Caricato, X. Li, H. P. Hratchian, A. F. Izmaylov, J. Bloino, G. Zheng, J. L. Sonnenberg, M. Hada, M. Ehara, K. Toyota, R. Fukuda, J. Hasegawa, M. Ishida, T. Nakajima, Y. Honda, O. Kitao, H. Nakai, T. Vreven, J. A. Montgomery, J. E. Peralta, F. Ogliaro, M. Bearpark, J. J. Heyd, E. Brothers, K. N. Kudin, V. N. Staroverov, R. Kobayashi, J. Normand, K. Raghavachari, A. Rendell, J. C. Burant, S. S.

Iyengar, J. Tomasi, M. Cossi, N. Rega, J. M. Millam, M. Klene, J. E. Knox, J. B. Cross, V. Bakken, C. Adamo, J. Jaramillo, R. Gomperts, R. E. Stratmann, O. Yazyev, A. J. Austin, R. Cammi, C. Pomelli, J. W. Ochterski, R. L. Martin, K. Morokuma, V. G. Zakrzewski, G. A. Voth, P. Salvador, J. J. Dannenberg, S. Dapprich, A. D. Daniels, O. Farkas, J. B. Foresman, J. V. Ortiz, J. Cioslowski, D. J. Fox, Gaussian 09, Revision B.01. In *Gaussian 09, Revision B.01, Gaussian, Inc.*, Wallingford CT, 2009. [b] M. J. Frisch, G. W. Trucks, H. B. Schlegel, G. E. Scuseria, M. A. Robb, J. R. Cheeseman, J. A. Montgomery, Jr., T. Vreven, K. N. Kudin, J. C. Burant, J. M. Millam, S. S. Iyengar, J. Tomasi, V. Barone, B. Mennucci, M. Cossi, G. Scalmani, N. Rega, G. A. Petersson, H. Nakatsuji, M. Hada, M. Ehara, K. Toyota, R. Fukuda, J. Hasegawa, M. Ishida, T. Nakajima, Y. Honda, O. Kitao, H. Nakai, M. Klene, X. Li, J. E. Knox, H. P. Hratchian, J. B. Cross, V. Bakken, C. Adamo, J. Jaramillo, R. Gomperts, R. E. Stratmann, O. Yazyev, A. J. Austin, R. Cammi, C. Pomelli, J. W. Ochterski, P. Y. Ayala, K. Morokuma, G. A. Voth, P. Salvador, J. J. Dannenberg, V. G. Zakrzewski, S. Dapprich, A. D. Daniels, M. C. Strain, O. Farkas, D. K. Malick, A. D. Rabuck, K. Raghavachari, J. B. Foresman, J. V. Ortiz, Q. Cui, A. G. Baboul, S. Clifford, J. Cioslowski, B. B. Stefanov, G. Liu, A. Liashenko, P. Piskorz, I. Komaromi, R. L. Martin, D. J. Fox, T. Keith, M. A. Al-Laham, C. Y. Peng, A. Nanayakkara, M. Challacombe, P. M. W. Gill, B. Johnson, W. Chen, M. W. Wong, C. Gonzalez and J. A. Pople, *Gaussian 03, Revision C.02*, Gaussian, Inc., Wallingford CT, 2004.

[24] P. Y. Ayala, H. B. Schlegel, *J. Chem. Phys.* 108, 2314 (1998).

[25] J. P. Guthrie, *J. Phys. Chem. A* 105, 8495 (2001).

[26] S. H. Maron and C. F. Prutton, *Principles of Physical Chemistry, 4th Ed.*, The Macmillan Company, New York (1965).

[27] D. R. Stull, E. F. Westrum Jr., G. C. Sinke, *The Chemical Thermodynamics of Organic Compounds*, John Wiley & Sons, New York (1969).

[28] F. A. Cotton, *Chemical Applications of Group Theory, 3rd Ed.*, John Wiley & Sons, New York (2008).

[29] Von D. F. Shriver, P. W. Atkins, C. H. Longford, *Inorganic Chemistry,* Oxford University Press, Oxford (1990).

[30] D. G. Truhlar, *J. Comput. Chem.* 12, 266 1991.

[31] R. B. McClurg, R. C. Flagan, W. A. Goddard, *J. Chem. Phys.* 106, 6675 (1997).

[32] A. D. Becke, *Phys. Rev. A* 38, 3098 (1988).
[33] C. Lee, W. Yang, R. G. Parr, *Phys. Rev. B* 37, 785 (1988).
[34] W. J. Hehre, R. Ditchfield, J. A. Pople, *J. Chem. Phys.* 56, 2257 (1972).
[35] J. D. Dill, J. A. Pople, J. A., *J. Chem. Phys.* 62, 2921 (1975).
[36] Head-Gordon, J. A. Pople, M. J. Frisch, *Chem. Phys. Lett.* 153, 503 (1988).
[37] S. Sæbø and J. Almlöf, *Chem. Phys. Lett.* 154, 83 (1989).
[38] M. J. Frisch, M. Head-Gordon, J. A. Pople, *Chem. Phys. Lett.* 166, 275 (1990).
[39] M. J. Frisch, M. Head-Gordon, J. A. Pople, *Chem. Phys. Lett.* 166, 281 (1990).
[40] M. Head-Gordon, T. Head-Gordon, *Chem. Phys. Lett.* 220, 122 (1994).
[41] R. Dennington, T. Keith, J. Milliam, *GaussView, version 5*, Semichem Inc., Shawnee Mission, KS, 2009.
[42] K. B. Wiberg, M. A. Murcko, *J. Am. Chem. Soc.* 110, 8029 (1988).
[43] N. Cohen, *J. Phys. Chem.* 96, 9052 (1992).
[44] C. Peng and H. B. Schlegel, *Israel J. of Chem.* 33, 449 (1993).
[45] C. Peng, P. Y. Ayala, H. B. Schlegel and M. J. Frisch, *J. Comp. Chem.*, 17, 49 (1996).

In: Computational Chemistry
Editor: Daria Bove

ISBN: 978-1-63321-354-8

Chapter 7

STRATEGIES FOR DESIGN OF NEW ORGANIC MOLECULAR RECTIFIERS: CHEMICAL AND MOLECULAR-SIMULATION PERSPECTIVES

Morad M. El-Hendawy[1,2], Niall J. English[3] and Ahmed M. El-Nahas[4]

[1]Chemistry Department, Faculty of Science,
Kafrelsheikh University, Kafrelsheikh, Egypt
[2]Higher Institute of Engineering and Technology, Kafrelsheikh, Egypt
[3]The SEC Strategic Research Cluster and the Centre for Synthesis and Chemical Biology, School of Chemical and Bioprocess Engineering,
University College Dublin, Belfield, Dublin, Ireland
[4]Chemistry Department, Faculty of Science,
El-Menoufia University, Shebin El-Kom, Egypt

ABSTRACT

Despite many efforts to date, the application of molecular rectification has been, and remains, very limited vis-à-vis use of traditional silicon-based rectifiers. In order to render the latter a more realistic proposition and to deliver tangible progress in this respect, a greater understanding of fundamental underlying mechanisms from a chemical perspective becomes sine qua non; indeed, molecular simulation has much to offer such an understanding. The three most common kinds of molecular rectifiers composed of Donor-σ-Acceptor, Donor-π-

Acceptor or Donor-Acceptor molecular systems, and their chemistry plays a vital role through the synthetic versatility of the functionality of molecular rectifiers. We have reviewed substitution, conformation, molecular length, anchoring, annulation, and isomeric effects as most common structural motifs used to design molecular rectifiers. The nature of the electrode and the external applied electric field also has a direct impact on rectification. The advent of experimental progress in molecular rectifiers has given rise new challenges to theory in developing theoretical tools to describe the electron transport in molecular rectifiers. Up to now, the combination of Density Functional Theory (DFT) and non-equilibrium Green's function (NEGF) methods have been widely used to study the quantum transport through nanoscale devices. Here, we offer a perspective on how molecular simulation approaches can be used to improve and inform design strategies for molecular rectifiers.

1. INTRODUCTION

Feynman was one of the first to predict a future for molecular-scale electronics. In a lecture in December 1959 at the annual meeting of the American Physical Society, entitled "*There is plenty of room at the bottom*", he described that the laws of physics do not limit our ability to manipulate single atoms and molecules: instead, it was the lack of appropriate methods for so doing so. Moore's insightful and provocative work on the growth of components into integrated circuits followed in the mid-1960's, leading to the eponymous 'law' in microchip design [1]. A pioneering breakthrough for molecular electronics came in 1974 by Aviram and Ratner, who introduced for the first time the possibility of functioning a single organic molecule as traditional diode for rectifying alternating current (AC) into direct current (DC) [2]. A schematic of an Aviram-Ratner (AR) rectifier is shown in Figure 1. DC is needed for the functioning of many electronic devices. DC is derived from AC using these types of AR-based rectifiers, which are composed of one or more diodes. An ideal diode is defined as a device which conducts only in one direction. There are two types of molecular diodes: rectifying diodes and resonating tunnelling diodes. In both of these devices, when the applied voltage exceeds a critical value, electrons are driven through one or more potential barriers. Since the 1990's, molecular electronics has become a fast-growing field of interest, with increasing levels of experimental and theoretical research activity.

However, in order for a molecular system to be considered as a device, there are several desiderata it must fulfil, in terms of functionality. A primary consideration of any molecular electronic candidate must be that of chemical stability, and it is important to understand the long-term stability of any molecular-electronics component under a wide variety of conditions. If a molecule tends to decompose when exposed to elevated temperatures, then it will not be a good candidate for use in a molecular electronic device. Similarly, the species must be inert with regards to other molecules of itself, and this would perhaps be the most obvious in cases of charge storage or redox-active molecules. Species which show such poor "insulation" would tend to exchange stored electrons, 'scrambling' any data represented by the storage of those electrons. In addition, a molecule which shows irreversible electron transfer would not be a good candidate for any sort of molecular-electronic device; indeed, the whole 'point' of electronics is to utilise *controlled* charge transfer from one element to the next. Finally, describing a molecule doing some useful function does not make it a molecular electronic device *per se*: there must be a way to interact with the component, both on a microscopic level and through input from the macroscopic world. Thus, a molecular electronic must be able to exchange information, or transfer states to other molecular electronic devices, and it must be able to interface with components in the system that are not nanoscopic. Naturally, these requirements present formidable challenges, and there is research proceeding to address them.

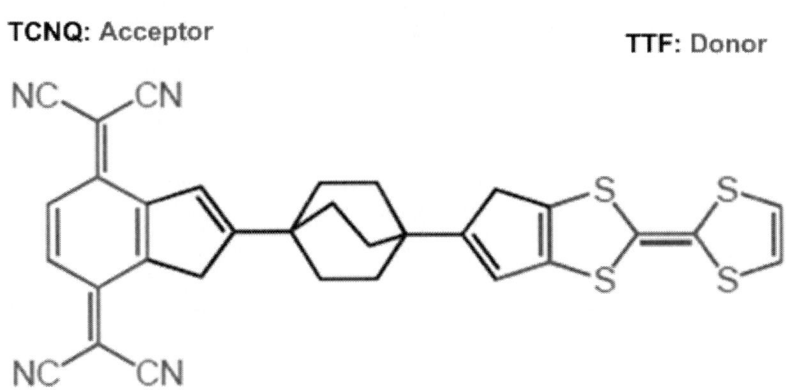

Figure 1. Molecular structure of an Aviram-Ratner rectifier (Donor-σ-Acceptor).

2. Classification and Mechanism

Classification Based on Molecular Structure

A molecular diode has two terminals and functions like a semiconductor *p-n* junction, as well as electronic states which can be clearly distinguished between highly conductive state (ON) and less conductive state (OFF). Based on the structure, the molecular rectifying diode can be classified into three types – *Donor-σ-Acceptor*, *Donor-π-Acceptor* and *Donor-Acceptor*, as illustrated in Figure 2.

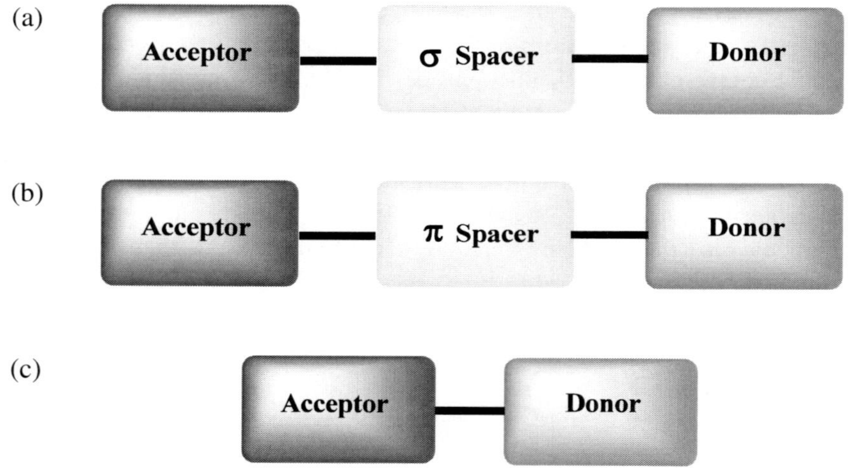

Figure 2. Classes of molecular diode based on their molecular structures: (a) Donor-σ-Acceptor, (b) Donor-π-Acceptor and (c) Donor-Acceptor.

A molecular resonance tunnelling diode (RTD) is based on a molecular conducting wire backbone [3]. For instance, two aliphatic methylene groups (CH_2) can be inserted on both sides of a benzene ring. Since aliphatic groups act as insulators, they create potential barriers to the flow of electrons in the molecular conducting wire. The only way for the current to flow in the presence of an applied voltage is when the electrons are forced to pass through the benzene. The operation of the molecular RTD can be understood in terms of potential barriers due to the two aliphatic methylene groups, the energy levels in the benzene ring through electron must traverse, and the orbital energy levels in the molecular conducting wire. In the presence of an applied voltage across the molecule, the molecular RTD will not conduct any current or will remain in OFF state since; the kinetic energy of the incoming electrons is different from those of the empty energy levels inside the benzene ring sandwiched between the two thin potential barriers. However, by changing the applied voltage, resonance can be induced, i.e., the kinetic energy of the incoming electrons can be made to match one of the unoccupied energy levels inside the benzene ring. The device will turn ON since electrons can traverse through the wire.

Mechanism of Rectification

Rectification is an electronic process in current flows preferentially in one direction. This can be examined by recording the current-voltage characteristics. Rectification of current is an essential parameter in fabricating any device for electronic purposes. The suitability of a particular molecule for use as diode is examined by determination of the rectification ratio, *R(V)*, the ratio of the current at two equal but opposite voltages:

$$R(V) = (I(V)_{forward} \,/\, I(-V)_{reverse}$$

Conventional semiconductor diodes exhibit rectification ratio from few hundreds depending on the doping materials [3]. R values can be significantly reached to approximately 10^7 as in Schottky-type diodes [3]. These values need to be achievable by molecular rectifiers but they depend on the nature of molecule, electrode material rather than the preparation process.

Aviram and Ratner proposed that the single asymmetric organic molecule can produce diode-like behaviour [1]: it was thought that electrons would flow only in one direction in a molecular system. As an ideal candidate they suggested a molecule containing a strong electron acceptor (as a p-type), tetracyanoquinodimethane (TCNQ), and a strong electron donor (as an n-type), tetrathiafulvalene (TTF), separated by a saturated σ-bridge to avoid coupling the molecular orbitals of the donor and acceptor subunits (cf. Figure 1). A junction is formed by bridging such a molecule with two electrodes forming semiconductor-like *p-n* junction. This molecule has never been synthesised, and thus experimental verification of its potential rectifying behaviour is difficult to estimate. However, there has been some estimation of the energies required for the molecule to undergo charge transfer from the ground state to the first excited state, and therefore rectify the current.

According to Metzger [4], this energy is about 3.5 eV, whereas the reverse process would need an energy of 9.6 eV (calculations were made for the ions at infinite separation).

In a recent patent, Ellenbogen and Love have proposed a systematic representation of the mono-molecular rectifying diode [5]. This diode is based on a molecular conducting wire consisting of two identical sections (S1, S2) separated by an insulating group R. Section S1 is doped by at least one electron donating group (X) = (e.g. $-NH_2$, - OH, $-CH_3$, $-CH_2CH_3$, etc.) and section S2 is doped by at least one electron withdrawing group (Y) = (e.g. $-NO_2$, -CN, -CHO, etc.). The insulating group (R) = (e.g. $-CH_2-$, $-CH_2CH_2-$) can be put into the molecular wire by bonding a saturated aliphatic group (with no π-orbitals). To adjust the voltage drop across R, multiple donor/acceptor sites can be incorporated. The single molecule ends are connected to the contact electrodes, e.g., gold.

To explain rectification of the mono-molecular diode under zero-bias, there are three potential barriers - one corresponding to the insulating group (middle barrier) and two corresponding to the contact between the molecule and the electrode (left and right barriers) [3, 5]. These potential barriers provide the required isolation between various parts of the structure. On the left of the central barrier, all the π-type energy levels (HOMO as well as LUMO) are elevated due to the presence of the electron-donating group, X, and, similarly on the right of the central barrier, energy levels are lowered due to the presence of the electron-withdrawing group, Y. This causes a built-in potential to develop across the barrier represented by the energy difference ΔE_{LUMO}. For current to flow, electrons must overcome the potential barrier from electron acceptor-doped section (S2) to the electron donor-doped section (S1). This forms the basis for the formation of the mono-molecular rectifying diode.

3. Experimental Observations for Molecular Rectifying Diodes

Research in molecular rectifiers has been focused on molecules, which can exhibit highly asymmetric I-V characteristics, at small values of applied voltage. Therefore, the creation of highly conducting molecular diodes requiring low operating powers should, hopefully, be possible. Here, we present some successful experimental attempts for a molecular systems exhibit rectification for the alternation current.

Aviram et al. observed asymmetrical I-V characteristics from a self-assembled monolayer (SAM) of quinone-based molecule deposited on gold (cf. Figure 3) [6]. Under negative bias, electron flowed from the tip to the quinone (acceptor) and from the catechol (donor) to the gold surface. However, soon after these results were published, the authors issued an erratum concluding that the rectifying behaviour may have been due to proton transfer from the catechol to the quinone. This process produces a semi-quinone structure, which would be a conductor, thereby resulting in enhanced current flow through molecule.

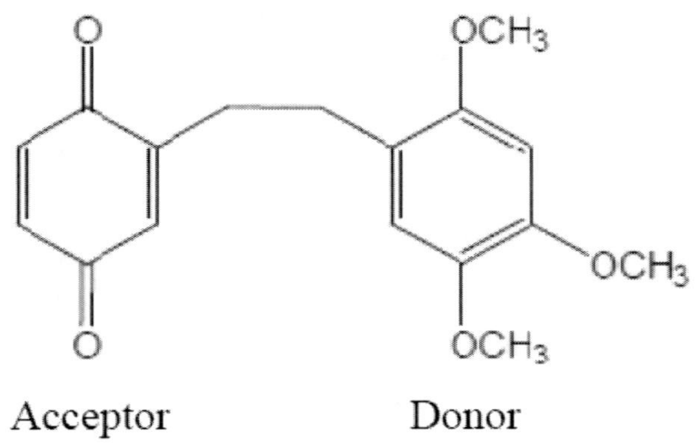

Figure 3. Aviram-Ratner Donor-σ-Acceptor molecule, based on the quinone acceptor and catechol donor [6].

There have been many attempts at the synthesis of Donor-σ-Acceptor molecules for potential unimolecular rectifiers, and of their investigation [3]. Mikayama et al. [7] synthesised a molecule containing a dinitrobenzene acceptor and a dihydrophenazine donor separated by a methylene σ-bridge (cf. Figure 4). The I-V curves displayed rectification ratio of ~ 7 at ±1 V under illumination, and *R(V)* of ~ 3 at ±1 V in the dark.

For rectification from Donor-π-Acceptor molecules, the donor subunit is insulated by π-spacer than the acceptor subunit. γ-hexadecylquinolinium tricyanoquinodimethanide ($C_{16}H_{33}$-Q3CNQ), has been sandwiched by Metzger et al. [8] between aluminum and magnesium electrodes; to prevent the formation of a Schottky barrier, Mg was covered by additional layer of Al (cf. Figure 5).

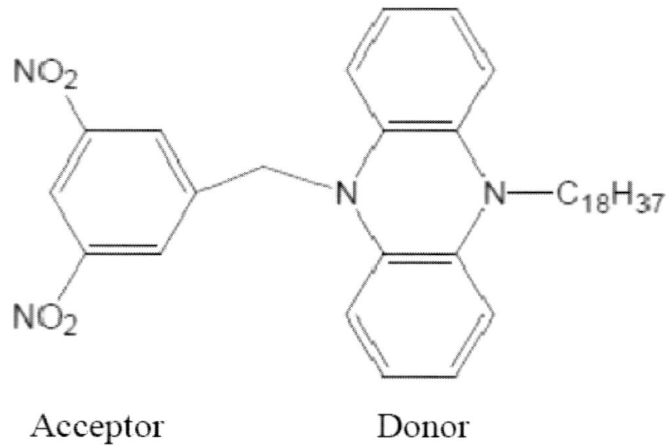

Figure 4. Chemical structure of D-σ-A molecule, investigated by Mikayama [7].

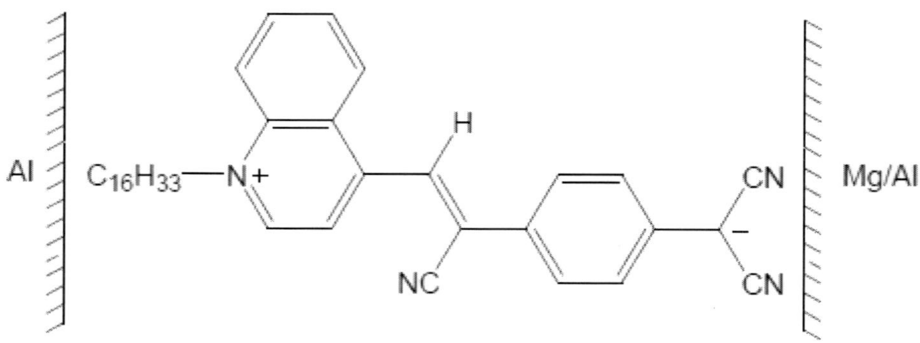

Figure 5. Molecular structure of D-π-A $C_{16}H_{33}$-Q3CNQ rectifier presented in a charge-separated form, placed between metallic electrodes [8].

Metzger et al. observed substantial asymmetry in the positive quadrant of the I-V plot, suggesting electron flow from 3CNQ to the quinolinium part [8], which led them to suggest the following mechanism of electron transfer:

$$M_1|D^+ - \pi - A^-|M_2 \rightarrow M_1|D - \pi - A|M_2 \quad (1)$$

$$M_1|D - \pi - A|M_2 \rightarrow M_1^+|D^+ - \pi - A^-|M_2^- \quad (2)$$

Eqn. 1 indicates intramolecular electron transfer by tunnelling through the π-bridge, whilst eqn. 2 represents transfer from D to M1 and from M2 to A. This behaviour was assigned to the reverse of Aviram-Ratner electron transfer through a molecule, in which 3CNQ acts as an acceptor, and quinolinium acts as a donor. According to theoretical calculations, $C_{16}H_{33}$-Q3CNQ molecule exists as a zwitterion in the ground state whilst in the excited state it adapts the quinonoid (neutral) form. Reproducible electric results gave a maximum rectification ratio of *ca.* 20 at ±1.5 V for a monolayer, and *ca.* 5 at ± 2.0 V for multilayer films.

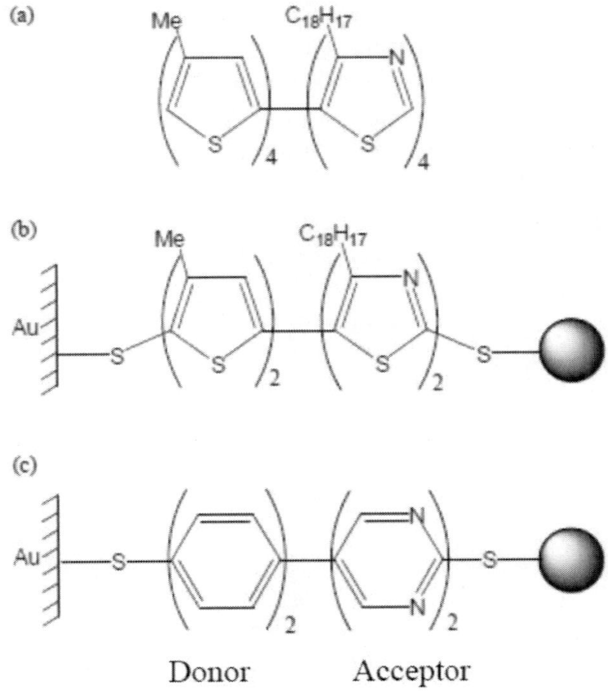

Figure 6. Diblock oligomers of Yu et al. [12] showing rectifying behaviour in (a) *Langmuir*-Blodgett (LB) films with a rectification ratio of 13 at ±1.5V, and SAMs in contact with Au nanoparticles with a rectification ratio of (b) 5 at ±1.5 V; (c) 4.5-9 at ±1.5 V.

In the case of rectification from diblock molecules, Ellenbogen and Love first suggested that linear conjugated oligomers can easily exhibit properties of *p-n* junctions by precise substitution [9]. Subsequently, Yu et al. carried out numerous experiments with diblock molecular diodes assembled on gold [10-

12]. Some of the investigated systems are presented in Figure 6. All of these systems contained electron-rich and electron-poor segments, corresponding to donor and acceptor moieties, respectively. Their asymmetric behaviour was rather small and resulted from the weak electron properties of the molecule system. The difference in rectification ratios between the first two molecules may well have resulted from different characteristics of the forces 'binding' the molecules to the substrate electrode. The rectification ratio was almost twice as high for the third molecule vis-à-vis the middle one, which was attributed to differences in the respective theoretically calculated dipole moments of 6.3 and 1.6 D.

In terms of rectification from mixed monolayers, Ashwell and Berry incorporated two types of deposition to obtain rectification [13], and a schematic is shown in Figure 7. Self-assembly was used to deposit a monolayer of asymmetric squaraine on a gold-coated substrate, and the *Langmuir*-Blodgett (LB) method used to form a second layer of symmetric squaraine. STM analysis indicated asymmetric I-V characteristics, which did not exhibit rectifying behaviour for both molecules when studied individually; this was attributed to the donor-acceptor interaction between the two layers in the hybrid system.

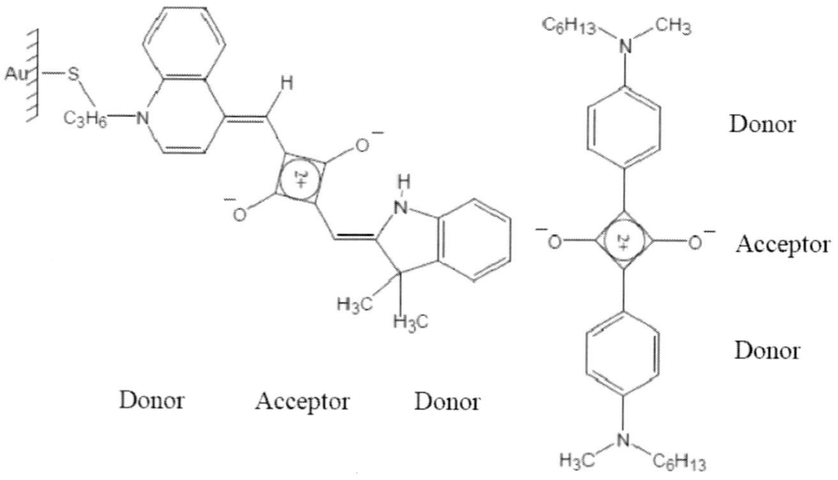

Figure 7. Molecular structure of hybrid SAM/LB rectifier, investigated by Ashwell and Berry [13].

Beyond the use of mixed monolayers, the recent advance of using few-layer nanogaps of graphene in the design of single-molecule diodes is an exciting advance [14, 15], and theoretical and simulation approaches have a good deal to contribute to this. In section 4, we consider the contribution of theoretical and simulation studies towards diodes.

4. THEORETICAL AND SIMULATION ASPECTS

Although the electrical properties of molecular devices with a finite number of molecules between metallic electrodes have been measured experimentally (*vide supra*); it is often difficult to interpret all of the observed properties with acuity. As mentioned earlier, earlier theoretical treatment considered a device based on a single organic molecule, on the Aviram-Ratner model [1]. Subsequently, much attention was devoted by many groups in the world towards incorporating different molecules. In this context, we will review theoretical efforts in respect to molecular rectifying diodes, especially the recent class of diblock diodes (D-A).

In general, the electrical properties of molecules can be calculated by either or both of the quantum chemical or electron transport methods. The electron transport method uses a two-probe configuration, wherein the molecule is sandwiched between two electrodes, to calculate experimental observable quantities like current or conductivity The quantum chemical method deals with the studying of electronic structures in the absence and presence of an electric field. Some authors prefer to calculate the effect of electric field on isolated molecules because they consider electrical properties of a given molecular device as an intrinsic character of the molecule itself. Others simulate the effect of electrodes by connecting the molecule with two small clusters representing the metal as semi-infinite electrodes. On the other hand, some recent studies tried to combine both approaches them to validate their interpretations. Kawazoe and co-workers have made contributions in the design of new candidates of molecular rectifiers through mainly studying the substituent effect on the geometry and electronic structure of a group of D-σ-A molecules. For example, they studied D-spacer-A molecular systems based on mono- or disubstituted benzene rings, covalently linked by an insulating bridge (methylene or dimethylene) [16]. The molecular structure possesses —SH groups for chemical grafting and easy electron injection from electrode to substrate. The individual donor and acceptor subunits have been designed by substituting of —NH_2 and —NO_2, groups. Calculation of molecular orbital energy levels and spatial orientation of the unoccupied molecular orbitals showed that in this kind of molecule, the lowest unoccupied molecular orbital (LUMO) was always localised on the acceptor ring, whereas the highest occupied molecular orbital (HOMO) was on the donor ring. However, the localisation of the LUMO on the donor side depends strongly on the number of substituents and the length of the bridge. The potential drop across the tunnelling barrier for a monosubstituted molecule with a methylene bridge was 1.56 eV, whereas for a disubstituted molecule with a dimethylene bridge it was 2.76 eV [16].

Rao studied the electron transport abilities for a series of donor/acceptor-heterocycles connected to the cumulenic bridge qualitatively using the ab initio Frontier Molecular Orbital (FMO) calculations [17]. FMO analysis showed that molecules having one sp-carbon atom in the cumulenic bridge act as rectifiers and can be treated as ideal alternatives to the traditional D-σ-A type of molecular rectifiers, whereas other molecules with the two sp-carbon atom behave as conductors and can be used as alternatives to the molecular wires in molecular devices. Furthermore, to support the FMO analysis for the conductance behaviour of the latter molecules, analysis has been carried out using the semi empirical Extended Hückel Theory and Non-Equilibrium Green Function (EHT-NEGF) formalism, which confirms the FMO results of the proposed molecular conductors [17].

Staykov et al. investigated the electrical rectifying properties of D-π-A system based on a single-molecule nanowire by means of the NEGF, combined with DFT (DFT-NEGF) [18]. An oligo-1,4-phenylene ethylene nanowire with π-donor and π-acceptor groups was attached on opposite sides of the molecule. The donor and acceptor wires were separated by a π-bridge, in contrast to the Aviram-Ratner rectifier, which is a D-σ-A diode. A model more similar to the real molecular electronic device is considered with relaxation of the molecular geometry, under the interaction with external electric field, taking into account its influence on the electronic properties of the nanowire. An asymmetric current-bias (I-V) diagram was observed, with a conductance ratio of 7. The analysis of the spatial distribution of FMOs, E_g, and the transmission spectra give an inside view of the observed results.

Recently, Dutta and Pati have examined computationally the electrical rectification of two isomeric organic molecules, naphthalene and azulene [19]. A single molecule of azulene showed a higher conductivity than naphthalene. The charge transfer from the seven-membered ring to the five-membered one in azulene gives the system its dipolar character. As a consequence, the conductance becomes asymmetric for forward and backward bias, opening the tantalising vista of efficient application of azulene as molecular rectifier in electronic-device applications.

The electric rectification of betaine-like donor-π bridge-acceptor systems has been demonstrated using a combination of *ab initio* techniques [20]. The results showed that electric rectification is extremely sensitive to the length of the chain, undergoing a complete switching after a specific chain length. This unique process is directly associated with a conjugated bridge in the presence of an external electric field. The conjugated bridge between the donor and acceptor groups is composed of oligoethylene with sizes ranging from 0 to 10 C=C units. The appearance of electric rectification occurs when the bridge size is equal to 5 units and is complete for those larger than 6 units (i.e., full inversion). This new electronic effect is advantageous for the design of large hybrid organic/inorganic circuits with an increased majority carrier flow that is necessary for the emerging needs of nanotechnology.

Classification Based on the Mechanism of Molecular Rectification

Unimolecular rectification based on voltage-controlled intramolecular stereochemical modification was suggested and computationally investigated [21]. In sharp contrast to present molecular rectifiers, these 'conformational' molecular rectifiers (CMRs) differ in principle from silicon structures, deriving their large, strongly temperature-dependent rectification from the differing current pathways in two dominant conformations, controlled by the inter-electrode applied field. Benzenethiol is a good example of a candidate for CMR [21]. In this molecule, the cyanomethyl group makes the dipole and this dipole can rotate under the action of applied field. When the CN- group points toward the tip, conductance is much higher. Because the molecular orbital of the CN- group is much nearer to the tip, electrons can transport between the tip and molecule [21]. The conductance will decrease when the CN group is pointing away from the contact. This behaviour showed rectification in the I-V curve; one situation has high conductance and by changing the voltage polarity, the other has lower conductance [21]. In addition, Seminario and co-workers have suggested a 'programmable' molecular diode driven by charge-induced conformational changes [22].

Quantum transport via a D-π-A single molecule (substituted p-terphenyl) was studied using DFT in conjunction with NEGF by He et al. [23]. Asymmetric electrical response for opposite biases was observed, resulting in significant rectification in current. The intrinsic dipole moment induced by substituent side groups in the molecule was found to leads to enhanced/reduced polarisation of the system under a forward/reverse applied potential, thus leading to asymmetry in the charge distribution and the electronic current under bias. Under a forward bias, the energy gap between the D and A frontier orbitals closes and the current increases rapidly, whereas under a reverse bias, the D-A gap widens and the current remains small.

Instead of asymmetric molecular substitution, molecular rectification could originate plausibly from asymmetrical electrode-molecule contacts. An example has been investigated is porphyrin-based molecular junction [24]: the asymmetry introduced by the addition of an extra thiol group on one side, against only one on the other side, was found to enhance the level of electronic coupling. The transportation of these contact-asymmetrical junctions showed obvious rectification, implying that the asymmetrical interface modification is feasible in the design of molecular diodes. Theoretical calculations using the DFT-NEGF method showed that the rectification ratio is about 2.6 in the large bias range from 0.6 to 1.2 V. This could be interpreted based on the alignment of the molecular orbital levels to the biased electrodes in the metal-molecule-metal junction [24]. These results highlight the fact that contact asymmetry is a significant factor to be considered when evaluating nanoelectrical junctions incorporating single molecules.

Recently, and after the experimental establishment of molecular rectification from diblock diodes, some computational groups have begun to pay significant attention to this subject in order to understand the nature of this rectification [25-32]. The electronic transport properties of polar conjugated molecules with different molecular lengths sandwiched between two gold electrodes were investigated using DFT-NEGF [27]. The calculated I–V curve showed obvious asymmetric features with increasing molecular length. The observed asymmetric characteristic resulted from the dissimilar shift of the perturbed molecular energy levels and the spatial asymmetry of the tunnelling wave functions under the external bias voltage. It was conjectured that the rectifying effect could be further enhanced by introducing an asymmetric coupling [27].

Protonation effects on electron transport through dipyrimidinyl-diphenyl molecular junctions have been studied theoretically based on DFT-NEFG [28]. Protonation was found to lead to enhancements in both conductance and rectification. However, the experimentally observed inversion in rectifying direction was not found in the calculations. The preferential current direction was always from the pyrimidinyl to the phenyl side. The results indicate that the protonation of the molecular wire is not the only reason of the rectification inversion [28].

Asymmetric current–voltage (I–V) curves were computed for a conjugated biphenyl–bipirimidine diblock oligomer with two thiol end groups sandwiched between Au(111) electrode surfaces [25]. The method was based on DFT and determined self-consistently the electronic structure of the molecule coupled to the gold electrodes with varied electrochemical potentials. Spatial asymmetry in the molecule was found to be *sine qua non* in generating the I–V asymmetry.

Another study has shown the influence of anchoring group of molecular diodes on rectifying behaviour [26]. Two molecular diodes with different anchoring groups, which are based on diblock co-oligomeric structures, have been synthesised and characterised. Scanning tunneling spectroscopy (STS) measurements revealed the correlation of rectifying effects in these molecular diodes with anchoring groups such as thiol and isocyanide. The combination of theoretical calculation and experimental results on these molecular diodes demonstrated that the rectifying effect could be affected by the nature of anchoring groups due to the bond dipoles at the interface and internal polarisation inside the molecules.

Theoretical investigations on the diblock molecular diode, thiophene–thiazole compound, have been carried out at the Hartree–Fock (HF) level by considering the interaction under external bias by Li et al. [29]. It was demonstrated that the electronic structures of this kind of diode are essentially different from those based on the Aviram-Ratner model in terms of the energy levels of the FMOs as well as their spatial distributions. The introduction of the external bias modified both the geometric and electronic structures. In particular, the spatial distributions of the frontier molecular orbitals were also shifted under external bias. Moreover, all of these features exhibited a strong dependence on the polarity of the applied bias due to the 'built-in' intrinsic molecular asymmetric structures, which could be used to interpret intuitively the asymmetrical current–voltage behaviour of molecules [29].

El-Hendawy et al. have carried out DFT-NEGF studies of the effects of conformational and constitutional isomerisation on diblock molecular diodes, as well as 'molecular engineering' by increasing the number of nitrogen atoms in 'sandwiched' molecular diodes [30-32]. It was found that there is some level of control over the electrical properties with such molecular engineering [32], while there is some promising level of control over conformational states of a candidate dipyrimidinyl-diphenyl dithiol molecular diode.

Díez-Pérez et al. have studied charge transport through symmetric tetraphenyl and non-symmetric diblock dipyrimidinyl diphenyl molecules covalently bound to two electrodes [33]. The orientation of the diblock was controlled through a selective deprotonation strategy, and a method in which the electrode–electrode distance was modulated unambiguously determines the current–voltage characteristics of the single-molecule device. The diblock molecule exhibited pronounced rectification behaviour compared with its homologous symmetric block, with current flowing from the dipyrimidinyl to the diphenyl moieties. This was interpreted in terms of localisation of the wave function of the hole ground state at one end of the diblock under the applied field. At a large forward current, the molecular diode was found to become unstable and quantum point contacts between the electrodes form. Although experimental results of this study do offer proof-of-principle of single-molecule diodes, the junction resistance was found to be greater than 10 MΩ [33], rendering practical applications very challenging.

Dual-Functional Device Combining Switch and Diode

Diarylethene has attracted much attention as a nanoscale switching unit of great promise in potential applications to various molecular devices [34]. Light-induced conductance switching of a single diarylethene molecule placed between two electrodes has been investigated using mechanically controllable break junctions (MCBJ) [35], scanning tunneling microscopy (STM) [36], and self-organised interconnection [37]. Theoretical approaches based on NEGF have been performed to obtain good insights into the switching properties of diarylethene [38–42]. Tsuji et al. have investigated the electrical current switching and rectifying properties of asymmetric diarylethenes incorporating two different heterocyclic five-membered rings with opposite electronic demands (cf. Figure 8), using the DFT-NEGF approach [43]. Electron transport through both the open- and closed-ring forms of the asymmetric diarylethenes has been dominated mainly by HOMO resonance. The ON/OFF ratios of the asymmetric diarylethenes are comparable to those of symmetric ones. In particular, Si-containing asymmetric diarylethenes showed a larger ON/OFF ratio. The transmission spectra of the O- and Si-containing asymmetric diarylethene showed a larger response to an external electric field, leading to a larger rectification ratio [43]. Based upon population analysis for the heteroatoms, electron transport from an electron-poor moiety (acceptor) to an electron-rich moiety (donor) is more favourable than that in the opposite direction. This rectifying diode-like behaviour is similar to that in the Aviram–Ratner molecular diode comprised of a Donor–σ–Acceptor structure. Asymmetric electron density plays an important role in the rectifying properties of a molecular diode. The electronegativity difference between heteroatoms was used as a measure of electronic asymmetry in a molecular diode with heteroatomic defects, and a linear relationship between the electronegativity difference and maximum rectification ratio was obtained [43]. These finding suggest that O- and Si-containing asymmetric diarylethene should serve as promising putative candidate for consideration as a novel molecular electronic device combining a switch and a diode.

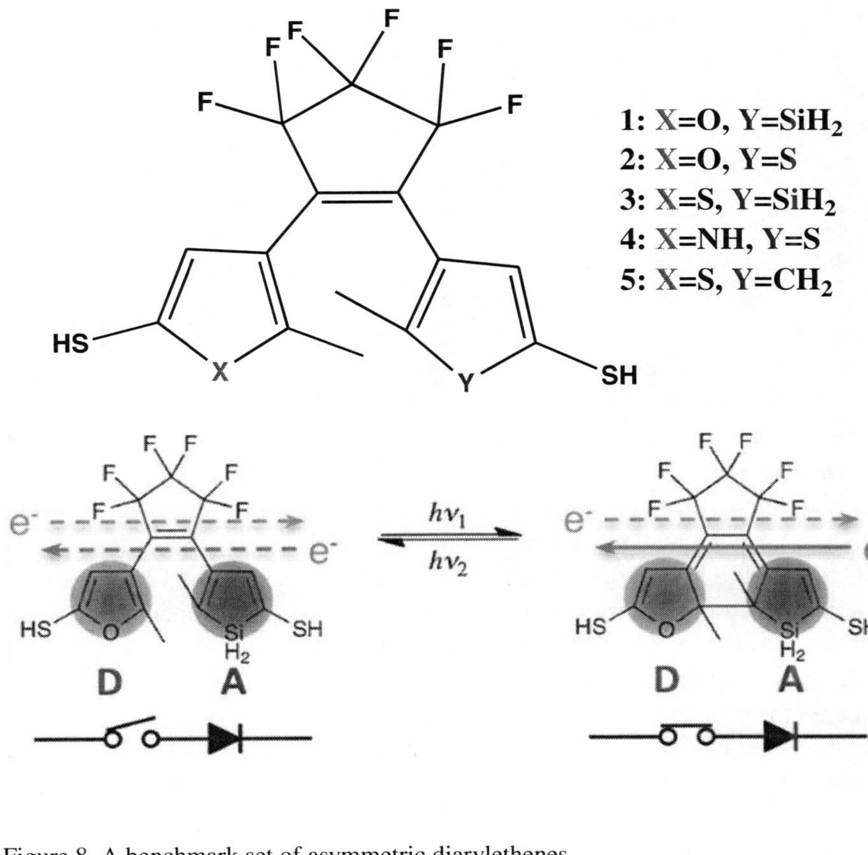

Figure 8. A benchmark set of asymmetric diarylethenes.

Effects of Conformation

To consider further the effects of conformation upon potential diodes, conjugated molecules have been investigated extensively over recent years. Usually, conjugated molecules consist of a rigid structure which makes it easier to predict which actual conformation is being investigated [44]. However, the conjugation itself depends on the overlap of the π-orbitals, which can be correlated directly with the torsion angle Φ between the individual π-systems. The first correlation between the conjugation and the conductance was demonstrated by Venkataraman et al. [45]. The compounds investigated were limited in their rotation around the biphenyl sigma bond; when the torsion angle was about 90°, low conductance was observed and when the torsion angle Φ was about 0° high conductance was observed. The observed conductance correlated linearly to the cos^2 of the torsion angle Φ between the two phenyl rings which was interpreted as an increased level of decoupling of the individual π-systems of both phenyl rings with increasing torsion angle Φ in the molecule. The $cos^2(\Phi)$ relationship arises from the π-orbital overlap integral A_{RS}, which is proportional to the resonance integral β_{RS} (R and S are decoupled chromophores). The interaction between the two orbitals across the sigma bond is defined by the resonance energy E_{RS}, which is known to correlate approximately with $cos^2(\Phi)$ [46, 47]. The experimental approach to investigate this correlation was later improved by Vonlanthen et al. [48], who synthesised a series of biphenyl derivatives which were restricted in their rotation around the central sigma bond by bridging alkanes. The influence of the torsion angle on the conductance is not only applicable to small biphenyl systems but also to larger structures like OPE-type wires [49, 50]. It is known that the planarisation of conjugated π-system leads to a reduction of the HOMO-LUMO gap, which can be observed easily by a red-shift in the absorption spectrum [51]. The correlation between the torsion angle and the conductance has motivated many research groups to design and synthesise a 'molecular switch' wherein a single benzene unit could be triggered to rotate in- and out- of plane [52-55].

The effect of low conjugation through a twisted biphenyl has been used to decouple an electron-rich from an electron-poor system generating a donor-bridge-acceptor (D-σ-A) molecule [56, 57]. However, the proposed rectifier 1' contained an electron poor fluorinated benzene unit, an electron-rich benzene unit and a central biphenyl with two additional methyl substituents forcing the two phenyl rings out of plane [56]. When the bias voltage was increased continuously, the energy levels of both units shifted relative to each other [56].

Whenever an unoccupied level passed by an occupied one, an additional transport channel opened up and the current increased. This is a good example of the influence of the torsion angle Φ on the conjugation; the twist between the biphenyl rings was sufficient enough to decouple the donor from the acceptor part, to create a rectifier [56].

Anchor-Rectification Relationship

The nature of the contact between the molecule and the electrode is an important characteristic of the junction. Ideally, a low-barrier ohmic contact between the metal and organic molecules will allow one to study 'pure' molecular electronic behaviour. However, in real junctions, most interfaces used to study the conductivity of the molecules and molecular assemblies have significant barriers for electrons or hole injection, which may 'dominate' the whole junction's behaviour. This contact controls the energy and mixing between molecular orbitals and the electronic states of the metals [58]. It has been demonstrated that a conjugated molecular wire, chemically bound to one gold electrode by the thiol linker and only in physical contact with the second gold electrode shows current rectification [59]. The effect of the different contacts was shown in a series of experiments by keeping the contact on one end of the molecule constant (S-Au) and varying the contacts on the other end [59, 60]. Since one end of the junctions is always the Au-S, the observed rectification ratio can be related to the amount of electronic coupling between the molecule and the metal at the second contact. The more effective the interaction between the molecule and the metal, the more symmetric the I-V characteristics. However, poor orbital mixing at the metal-molecule interface results in strong current rectification. Thus, this underscores clearly the importance of metal/molecule contact in charge transport of the entire junction.

The nature of the metal-molecule junction can have substantial influence on the performance of the device. One of the important issues in regard to analysis of results are the difficulties associated with understanding the chemical nature and structure of the junctions and contacts that are being measured. The use of a wide range of current analytical techniques can provide exhaustive information about the chemical composition of the bulk materials and thin films, the alignment of the molecules within the monolayer, and the properties. There are also many approaches to fabricate the electrode-molecule-electrode junctions, such as vapour deposition of the top electrode on the monolayer and liquid electrode junctions that allow measurement of conductivity of a small number of molecules.

CONCLUSION

This review has offered perspectives on design strategies for novel organic molecular rectifiers, with particular emphasis from the vantage points of chemical and molecular-simulation aspects. It is important to bear in mind that the conceptual simplicity of the Aviram-Ratner design belies some formidable limitations, such as extreme sensitivity to energy-level alignment of molecular orbitals with each other and with connecting electrodes rendering experimentally predictable designs elusive, and the need for a σ bridge adding a large tunnel barrier to the backbone and leading to substantial junction resistance. As a result, an experiment that has demonstrated proof-of-principle of single-molecule diodes have resistances greater than 10 MΩ [31]. Key desiderata to further progress in this field are predicated on design principles involving 'tunability' of rectification in single-molecule diodes [30, 61], and DFT-NEGF has proven its utility as a useful design tool in this regard. It is to be hoped that the coming decade or so will see advances in massively-parallel and (near-) linear-scaling DFT and *ab initio* simulation techniques being able to design molecular diodes and their environments with ever-increasing accuracy, ushering in an era of genuinely predictive, simulation-led *in silico* design of molecular rectifiers, as a rapid design and prototyping tool.

REFERENCES

[1] G. E. Moore, *Electronics,* 38, 114-117 (1965)

[2] Aviram & M.A. Ratner, *Chem. Phys. Lett.* 29, 277-283 (1974).

[3] G. D. Goldhaber, M.S. Montemerlo, J.C. Love, G.J. Opiteck and J.C. Ellenbogen, *Proc IEEE* 85, 521-540 (1997).

[4] R. M. Metzger, *Ann. N.Y. Acad. Sci.* 252, 1006 (2003)

[5] J. C. Ellenbogen & J.C. Love, US6348700 B1 (1998)

[6] Aviram, C. Joachim and M. Pomerantz, *Chem. Phys. Lett.* 146, 490 (1988)

[7] T. Mikayama, M. Ara, K. Uehara, A. Sugimoto, K. Mizuno and N. Inoue, *Phys. Chem. Chem. Phys.* 3, 3459 – 3462 (2001)

[8] R. M. Metzger, B. Chen, U. Höpfner, M.V. Lakshmikantham, D. Vuillaume, T. Kawai, X. Wu, H. Tachibana, T.V. Hughes, H. Sakurai, J. W. Baldwin, C. Hosch, M.P. Cava, L. Brehmer & G.J. Ashwell, *J. Amer. Chem. Soc.* 119, 10455 – 10466 (1997)

[9] J. C. Ellenbogen and J.C. Love, *Proc. IEEE* 88, 386 – 426 (2000).

[10] M-. K. Ng & L. Yu, Angew. *Chem. Int. Ed.* 41, 3598 – 3601 (2002).

[11] P. Jiang, G.M. Morales, W. You & L. Yu, Angew. *Chem. Int.* Ed. 43, 4471 – 4475 (2004)

[12] G. M. Morales, P. Jiang, S. Yuan, Y. Lee, A. Sanchez, W. You & L. Yu, *J. Amer. Chem. Soc.* 127, 10456 – 10457 (2005).

[13] G. J. Ashwell and M. Berry, *J. Mater. Chem.* 15, 108-110 (2005).

[14] F. Prins, A. Barreiro, J.W. Ruitenberg, J.S. Seldenthuis, N. Aliaga-Alcalde, L.M.K. Vandersypen and H.S. van der Zant, *Nano. Lett.* 11, 4607-4611 (2011)

[15] Bergvall, K. Berland, P. Hyldgaard, S. Kubatkin and T. Löfwander, *Phys. Rev.* B 84, 155451 (2011)

[16] Majumder, H. Mizuseki and Y. Kawazoe, *J. Phys. Chem. A*, 105, 9454 – 9459 (2001).

[17] J. L. Rao, Central Euro. *J. Chem.* 5, 793-812 (2005)

[18] Staykov, D. Nozaki, K. Yoshizawa, *J. Phys. Chem. C,* 111, 11699 (2007)

[19] S. Dutta & S. K. Pati, *Resonance* 14, 80 (2008).

[20] Saraiva-Souza, B. G. Sumpter, V. Meunier, A.G. Souza, J. Del Nero, *J. Phys. Chem. C.* 112, 12008 (2008).

[21] Troisi and M.A. Ratner, *Nano Lett.,* 4 (2004) 591.

[22] P. A. Derosa, S. Guda, J. M. Seminario, *J. Am. Chem. Soc.* 125, 14240 (2003).

[23] H. He, R. Randey, G. Mallick, S. P. Karna, *J. Phys. Chem. C* 113, 1575 (2009).

[24] J. Zhao, C. Yu, N. Wang, H. Liu, *J. Phys. Chem. C* 114, 4135 (2010)

[25] Z. L. Gasyna, G. M. Morales, A. Sanchez, L.Yu, *Chem. Phys. Lett.,* 417 (2006) 401.

[26] Y. Lee, B. Carsten, L. Yu, *Langmuir,* 25 (2009) 1495.

[27] J. Huang, Q. Li, Z. Li, J. Yang, J. *Nanosci. Nanotechnol.,* 9 (2009) 774.

[28] L. Zhenyu, H. Jing, L. Qunxiang, Y. Jinlong, Sci *China Ser B-Chem*, 51 (2008) 1159.

[29] Y. W. Li, J. H. Yao, C. L. Yang, S. K. Zhong, G. P. Yin, *Mol. Simulat.,* 35 (2009) 301.

[30] M. M. El-Hendawy, A. M. El-Nahas, M. K. Awad, *J. Phys. Chem. C,* 114 (2010) 21728.
[31] M. M. El-Hendawy, A. M. El-Nahas, M. K. Awad, *Org. Electron.,* 12 (2011) 1080.
[32] M. M. El-Hendawy, A. M. El-Nahas, M. K. Awad, *Org. Electron.,* 13 (2012) 807-814.
[33] Díez-Pérez, J. Hihath, Y. Lee, L. P. Yu, L. Adamska, M.A. Kozhushner, I.I. Oleynik and N. J. Tao, *Nat. Chem*., 1 (2009) 635.
[34] S. J. van der Molen, P. Liljeroth, *J. Phys.: Condens. Matter* 2010, 22, 133001.
[35] D. Dulić, S. J. van der Molen, T. Kudernac, H. T. Jonkman, J. J. D. de Jong, T. N. Bowden, J. van Esch, B. L. Feringa, B. J. van Wees, *Phys. Rev. Lett.* 2003, 91, 207402.
[36] S.J. van der Molen, H. van der Vegte, T. Kudernac, I. Amin, B. L. Feringa, B. J. van Wees, *Nanotechnology* 2006, 17, 310.
[37] M. Taniguchi, Y. Nojima, K. Yokota, J. Terao, K. Sato, N. Kambe, T. Kawai, *J. Am. Chem. Soc*. 2006, 128, 15062.
[38] M. Kondo, T. Tada, K. Yoshizawa, Chem. Phys. Lett. 2005, 412, 55.
[39] Staykov, D. Nozaki, K. Yoshizawa, *J. Phys. Chem.* C 2007, 111, 3517.
[40] Y. Tsuji, A. Staykov, K. Yoshizawa, *Thin Solid Films* 2009, 518, 444.
[41] Y. Tsuji, A. Staykov, K. Yoshizawa, *J. Phys. Chem.* C 2009, 113, 21477.
[42] Odell, A. Delin, B. Johansson, I. Rungger, S. Sanvito, *ACS Nano* 2010, 4, 2635.
[43] Y. Tsuji, J. Koga, K. Yoshizawa, *Bull. Chem. Soc. Jpn.,* 86 (8) 2013, 947–954
[44] I.-W. P. Chen, M.-D. Fu, W.-H. Tseng, C. Chen, C.-M. Chou, T.-Y. Luh, *Chem. Commun.* 2007, 3074.
[45] L. Venkataraman, J. E. Klare, C. Nuckolls, M. S. Hybertsen, M. L. Steigerwald, *Nature* 2006, *442*, 904–907.
[46] H. H. Jaffé, M. Orchin, *Theory and Applications of Ultraviolet Spectroscopy*, Wiley-vch, 1962.
[47] D. Vonlanthen, Biphenyl-Cyclophanes: The Molecular Control over Conductivity of Single-Molecule Junctions, Basel, 2010.
[48] D. Vonlanthen, A. Mishchenko, M. Elbing, M. Neuburger, T. Wandlowski, M. Mayor, *Angew. Chem. Int. Ed.* 2009, *48*, 8886–8890.
[49] J. F. Smalley, S. B. Sachs, C. E. D. Chidsey, S. P. Dudek, H. D. Sikes, S. E. Creager, C. J. Yu, S. W. Feldberg, M. D. Newton, *J. Am. Chem. Soc.* 2004, *126*, 14620–14630.

[50] M. D. Newton, J. F. Smalley, *Phys. Chem. Chem. Phys.* 2007, *9*, 555.
[51] Tsuda, A. Osuka, *Science* 2001, *293*, 79–82.
[52] J. Chen, M. . Reed, *Chemical Physics* 2002, *281*, 127–145.
[53] J. Taylor, M. Brandbyge, K. Stokbro, *Phys. Rev. B* 2003, *68*, 121101.
[54] J. Chen, M. A. Reed, A. M. Rawlett, J. M. Tour, *Science* 1999, *286*, 1550–1552.
[55] Błaszczyk, M. Chadim, C. von Hänisch, M. Mayor, *Eur. J. Org. Chem.* 2006, 3809–3825.
[56] M. Elbing, R. Ochs, M. Koentopp, M. Fischer, C. Von Hänisch, F. Weigend, F. Evers, [55] H. B. Weber, M. Mayor, *PNAS* 2005, *102*, 8815–8820.
[57] R. M. Metzger, *Chem. Rev.* 2003, *103*, 3803–3834.
[58] Mujica, V.; Kemp, M.; Ratner, M. A., *J. Chem. Phys.* 1994, *101*, 6849-6855.
[59] Kushmerick, J. G.; Holt, D. B.; Yang, J. C.; Naciri, J.; Moore, M. H.; Shashidhar, R., *Phys. Rev. Lett.* 2002, *89*, 086802.
[60] Kushmerick, J. G.; Whitaker, C. M.; Pollack, S. K.; Schull, T. L.; Shashidhar, R., *Nanotech.* 2004, *15*, S489-S493.
[61] Batra, P. Darancet, Q. Chen, J.S. Meisner, J.R. Widawsky, J.B. Neaton, C. Nuckolls and L. Venkataraman, *Nano Lett.* 13, 6233–6237 (2013)

In: Computational Chemistry
Editor: Daria Bove
ISBN: 978-1-63321-354-8

Chapter 8

DFT STUDIES OF 8π,6π-ELECTROCYCLIZATIONS OF BENZOOCTATETRAENES AND BENZODECAPENTAENES

***Davor Margetić*[1], *Iva Jušinski*[1] *and Irena Škorić*[2]**
[1]Laboratory for Physical-Organic Chemistry,
Division of Organic Chemistry and Biochemistry,
Ruđer Bošković Institute, Bijenička cesta 54, 10001 Zagreb, Croatia,
[2]Department of Organic Chemistry, Faculty of Chemical Engineering and Technology, University of Zagreb, 10000 Zagreb, Croatia

ABSTRACT

The 8π,6π-electrocyclization processes of substituted conjugated tetraenes and pentaenes were investigated by quantum-chemical calculations - density functional (DFT) methods. Reactivity of substituted octatraenes with double bond incorporated into the benzene ring in which substituents are positioned at various position(s) of conjugated polyenes was studied. Influence of terminal substituents on reaction mechanism and the product *endo/exo* selectivity was also established.

Keywords: Pericyclic reactions, electrocyclization, octatetraenes, reaction mechanism, quantum-chemical calculations

1. INTRODUCTION - 8π,6π-ELECTROCYCLIZATIONS

Dienes and polyenes are organic compounds which are involved in a large number of organic reactions, and as such are constituents of a number of natural and human-made compounds, such as terpenes, cholesterol, and vitamin A. [1] Conjugated dienes and polyenes take place in their characteristic electrocyclization reactions, which are of great synthetic importance and believed to be involved in biosynthesis of many natural compounds. [2]

In this group of reactions, thermal 6π electrocyclizations of conjugated trienes are the most common of all electrocyclizations employed in the synthesis. [3] By such processes, all-carbon, as well as heterocyclic products could be formed. [4, 5] Conjugated cyclohexadiene units are the products which are thermodynamically favorable, but usually require activation by relatively high temperatures. The cyclohexadiene products formed by such process can be followed by further reactions such as Diels-Alder cycloadditions, and these reaction cascades efficiently form extremely complex molecular architectures. [6] Amongst them, the 8π,6π electrocyclic cascade (also referred to as Black's cascade) [7] is arguably the most elegant display of the power of electrocyclization reactions in the nature. This tandem reaction involves a cascade of two thermally induced, 8π, followed by 6π electrocyclizations leading directly to compounds that contain the bicyclo [4.2.0] octadiene ring system. [8]

The 8π electrocyclizations are commonly followed by a 6π electrocyclization, since the 8π reaction necessarily forms a triene with the *cis*-stereochemistry required for the 6π event. [9] 1,3,5,7-Octatetraenes in which the two internal olefins have the *Z*-configuration are thermally unstable with respect to the 8π,6π electrocyclization cascade, and tend to undergo further pericyclic reactions which could explain the scarce occurrence of this structural motif in natural products, and displays biosynthetic importance of 8π,6π electrocyclizations. [10, 11]

Biosynthetic origin of the endiandric acids by 8π,6π electrocyclization have been proposed by Black, [13] while Parker [14], Baldwin [15, 16] and Trauner [17] proposed 8π,6π electrocyclization mechanism in the synthesis of immunosuppressants SNF4435 C/D, which was later verified by total synthesis by Nicolaou (Scheme 1). [18, 19] More recently, Rickards proposed electrocyclase enzymes to support 6π-electron electrocyclic reactions *in vivo*. [20]

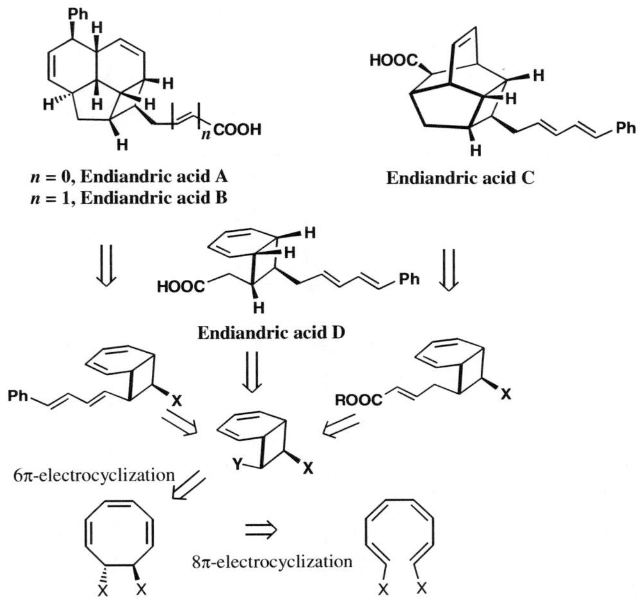

Scheme 1. Structures and retrosynthetic analysis of Endiandric acids A-D [12].

Several thermolysis of cyclooctatraene derivatives which follow the 8π,6π electrocyclization reaction mechanism has been reported in the literature, and the configuration around the double bonds was found to be of crucial importance for their reactivity. [21, 22] Examples are provided by compounds **1** [23] and **2** [15, 16] (Figure 1). In these molecules substituents are attached at the *termini* of tetraene moiety and 8π,6π-electrocyclization readily takes place. On the other hand, the 6π-electrocyclization of tetraenes **3** [24] and **4** [25] does not follow initial 8π-electrocyclization process. Similarly, there are examples of the 6π-electrocyclization which do not occur in octatetraene derivatives with the C=C bonds embedded into dihydropyran [26] rings.

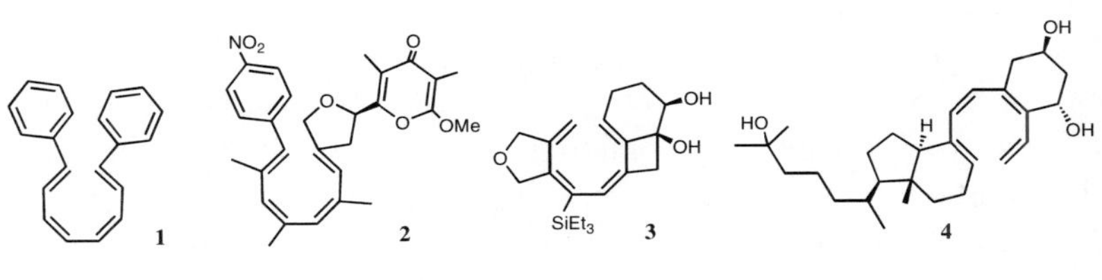

Figure 1. Literature examples of octatetraene systems undergoing 8π,6π electrocyclizations.

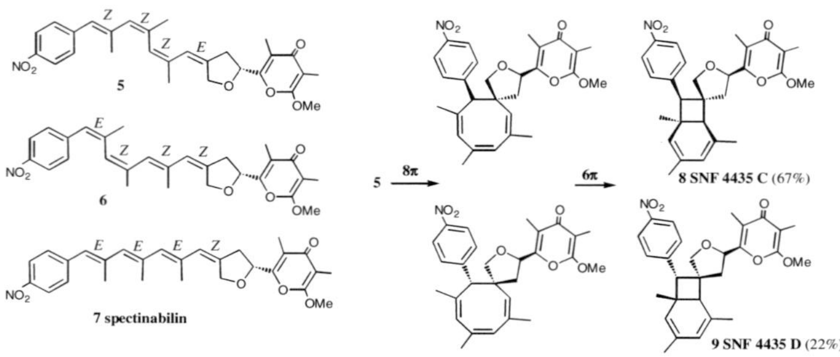

Scheme 2. Total syntheses of SNF4435 C and D through 8π,6π electrocyclization cascades.

Natural products that arise from 8π,6π electrocyclizations represented by SNF4435 C, ocellapyrones A and B [27], elysiapyrones A and B [28], or γ-pyrone from sacoglassan *Planobranchus ocellatus* [29, 30] have sparked significant synthetic activity, which has led to a better understanding of electrocyclization cascades and polyene chemistry in general and preparation of natural products such as shimalactones, [31] and pre-kingianin A [32]. The bicylo[4.2.0]octadiene core of natural products presumably arises biosynthetically through a thermal 8π,6π electrocyclization cascade from a highly substituted conjugated polyene. In compliance with the conrotatory nature of thermal 8π electrocyclizations, this would require a (*Z,Z,Z,E*)-configured precursor **5** or its (*E,Z,Z,Z*)- counterpart **6** (Scheme 2). These tetraenes are geometrical isomers of the antibiotic spectinabilin **7**, whose spontaneous or enzyme-catalyzed isomerization could lead for the formation of the SNF compounds **8** and **9**. [33]

The ratio of synthetic **SNF4435 C** and **D** formed in these cascades (3.0-3.8:1.0) closely parallels that of the compounds found in nature. The origin of this diastereoselectivity lies in the conrotatory 8π-electrocyclization of tetraene **6**, which can proceed through either of the helical transition states **10** or **12**. The former is slightly favored, because in this arrangement the bulky pyrone substituent is orientated away from the interior of the helix (Scheme 3).

In electrocyclic reactions conjugated *polyene* interconverts with an unsaturated cyclic compound containing one carbon-carbon double bond less than the starting polyene. These reactions can be promoted thermally and proceed with a very high degree of stereoselectivity, which could be interpreted by the orbital symmetry. [34] According to the Woodward-Hoffmann selection rules, [35] 8π-electrocyclization takes place *via* thermally allowed conrotatory ring closure, while the concomitant 6π-electrocyclization takes place by disrotatory mode (Scheme 4). In the case of unsubstituted cycloocatetraene (X=Y=H), only one bicylo[4.2.0]octadiene product is formed. Required geometry of the terminal double bonds of (*E,Z,Z,E*)-octatetraene for correct orbital overlap for 8π electrocyclization is depicted in Figure 2, while optimized transition state structures for 8π and 6π-electrocyclizations of archetypal (*E,Z,Z,E*)-octatetraene optimized by the M06-2X/6-31G* method are shown in Figure 3.

The 6π- and 8π-electrocyclizations of substituted octatetraenes were studied computationally employing a variety of quantum-chemical methods including PM3, [36] MINDO/3, [37] 3-21G, [38] 6-31G(d), [39] B3LYP/6-31G(d) [40] and MP2-4/6-31G* [48] methods.

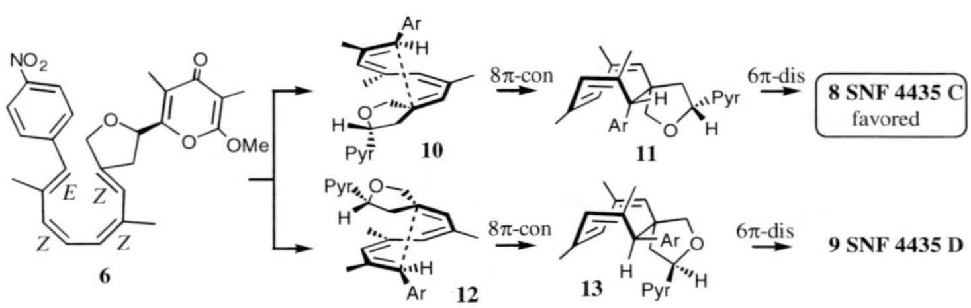

Scheme 3. Orientation of terminal substituents for 8π electrocyclization transition state structures in syntheses of SNF4435 C and D.

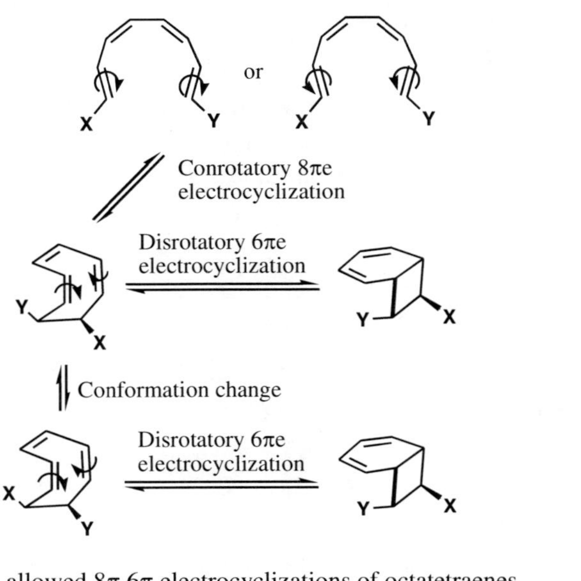

Scheme 4. Thermally allowed 8π,6π electrocyclizations of octatetraenes.

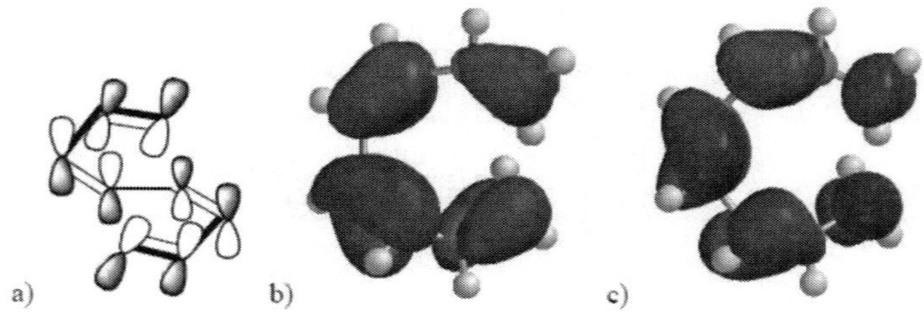

Figure 2. a) Required geometry of the terminal double bonds of (*E,Z,Z,E*)-octatetraene for correct orbital overlap for 8π electrocyclization, b) depiction of the HOMO orbital (RHF/6-31G*), and c) LUMO orbital.

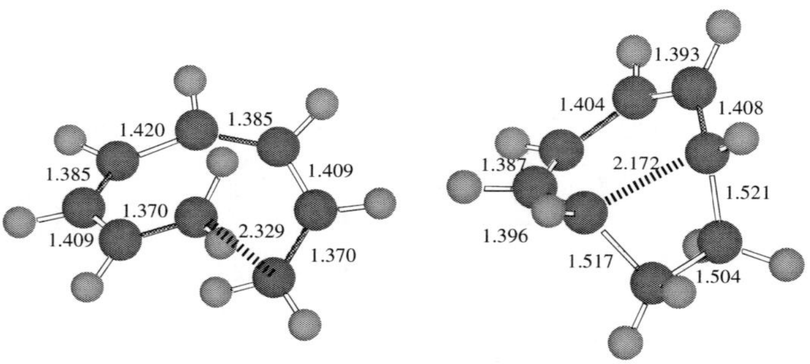

Figure 3. Transition state structures for 8π and 6π- electrocyclizations of (*E,Z,Z,E*)-octatetraene optimized by the M06-2X/6-31G* method (bond distances are given in Å).

On the othe hand, prior to our studies, domino 8π,6π-electrocyclizations were much less explored computationally, as far as we are aware, only Schreiner reported computational study using B3PW91/6-31G(d,p) method. [41]

2. Consequences of Incorporation of the Octatetraene π-System in the Aromatic Ring on the 8π,6π- Electrocyclizations of [3,4]-Benzo-(*E,Z,Z,E*)-1,3,5,7-Octatetraenes [42]

Computational investigation of 8π,6π- electrocyclizations of *o*-butadienylstyrenes, where conjugated tetraene π-system is at the same time part of an aromatic moiety pinpointed the consequences of incorporation of the octatetraene π-system in the aromatic ring on the mechanism of 8π,6π-electrocyclizations, and experimentally observed selectivities (Scheme 5). Results of thermal 8π,6π- electrocyclizations of *cis,trans*-benzooctatetraenes **14a-c** showed that incorporation of octatetraene π-system in the aromatic ring requires significantly higher temperatures for cyclization to take place than octatetraenes substituted at the termini. In addition, depending on the substituents, different ratios of two isomeric products, *endo*- and *exo*-7-substituted-2,3-benzobicyclo[4.2.0]octa-2,4-diene **15** were obtained. In the case of methyl substituent, mixture of *endo*-**15a** and *exo*-**15b** was achieved, while the substitution with either phenyl or 2-furyl group led to the formation of a single isomers (*endo*-**15b** and *endo*-**15c**).

Identical stereospecific 8π,6π-electrocyclic ring closures of fully (*E,E,E*)-conjugated skeleton of sulfonyl substituted benzotetraenes **16** to generate bicyclic products **17** were recently reported by Chang (Scheme 6). [43]

The transformation of benzooctatetraenes **14a-c** to *endo*- and *exo*-7-substituted-2,3-benzobicyclo[4.2.0]octa-2,4-dienes **15a-c** takes place *via* an 8π,6π- electrocyclization cascade, as shown in Scheme 7. In the first step (8π-electrocyclization) conrotatory ring closure *via* transition states **18a-c-TS** takes place. [44]

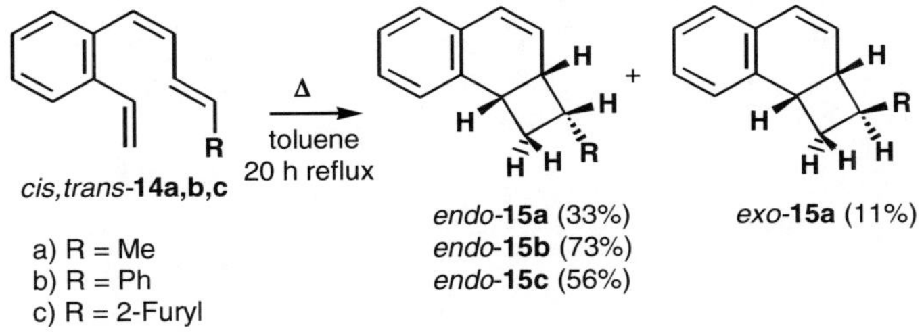

Scheme 5. Thermal reaction of benzooctatetraenes **14a-c.**

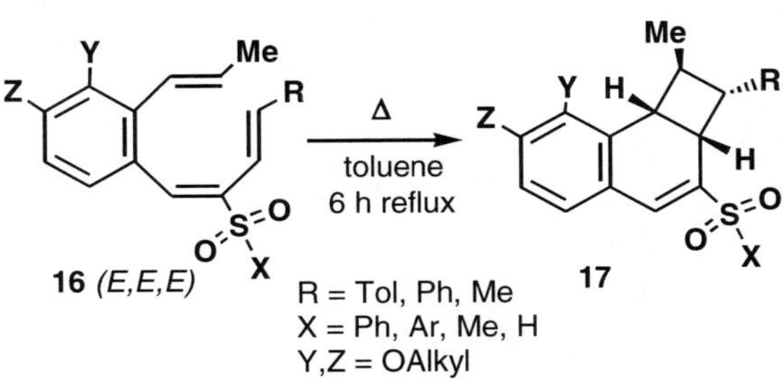

Scheme 6. Thermal reaction of benzooctatetraenes **16.**

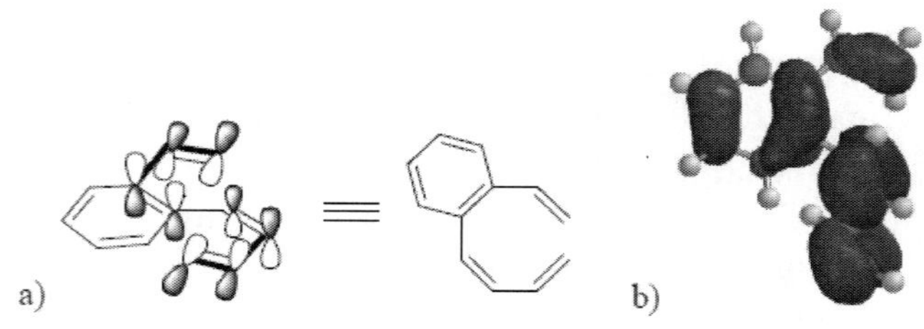

Figure 4. a) Required geometry of the terminal double bonds of benzooctatetraenes for correct orbital overlap for 8π electrocyclization and b) depiction of the HOMO orbital (RHF/6-31G*).

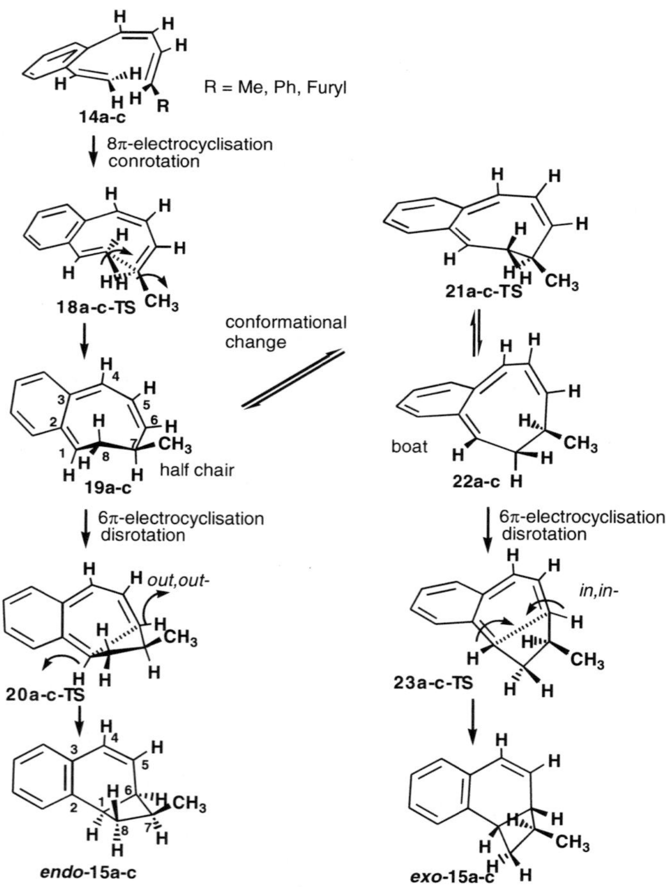

Scheme 7. Mechanism and stereochemical considerations for thermal electrocyclization of **14a-c**.

Required geometry of the terminal double bonds of benzooctatetraenes for correct orbital overlap is depicted in Figure 4. Final stereochemistry at C7 and C8 is defined at this point, leading to intermediate pentaenes **19a-c**. Benzene moiety is restored in the second reaction step (disrotatory 6π-electrocyclization), *via* transition states **20a-c-TS**. This process defines configurations at C1 and C6 atoms of *endo*-**15a-c**. To obtain *exo*-**8a-c** products, conformational interconversion (cyclohexatriene ring flipping of the CH_2 group) of the intermediates **19a-c** (half chair conformation) takes place *via* transition states **21a-c-TS** to boat conformers **22a-c**. In analogy to the intermediates **19a-c**, disrotatory 6π-electrocyclization of intermediates **22a-c** involving transition states **23a-c-TS**, gives the products *exo*-**15a-c**. Thermally orbital symmetry forbidden [2+2] cyclization and 6π-electrocyclization as initial reactions are predicted by the M06/6-31+G(d,p) calculations to be much higher in energy (221.2, and 174.4 kJ/mol, respectively), which rules out the possibility of these alternative reaction pathways.

Quantum-chemical study carried out at the M06/6-311+G(d,p)//M06/6-31+G(d,p) level was used to rationalize experimentally obtained stereospecifities. This M06 is novel hybrid meta functional with good accuracy for applications involving main group thermochemistry, kinetics and barrier heights. [45] Employment of this functional gained in recent times wide popularity amongst computational chemists since these calculations provide a more fair comparison of the energetics, as recent calculations have established that M06 leads to activation energies which are by a few kcal/mol more accurate than B3LYP for the pericyclic reactions. [46, 47]

The M06/6-31G+(d,p) calculated geometry of **18a-TS** (Figure 5) qualitatively resembles the B3LYP3/6-31G(d,p) optimised transition state structures for conrotatory 8π-electrocyclization of a series of octa-1,3,5,7-tetraenes published by Cossio. [40] These 'torquselective' helical transition state structures are analogous to those published by Houk and co-workers [40] of the 8π-conrotatory electrocyclizations of 1-substituted 1,3,5,7-octatetraenes. The most characteristic structural features of transition state **18a-TS** is the length of new forming C7···C8 bond (2.096 Å) and its helical conformation which allows excellent eclipsing between the terminal C1 and C8 atoms (Figure 5). The alternation of double bonds of the starting tetraene is partially preserved in **18a-TS** (Δr=0.070 Å). Furthermore, the elongation of the new forming C1···C6 bond in **20a-TS** (2.434 Å) and **23a-TS** (2.482 Å), as compared with **18a-TS** could be assigned to the geometry constraint present in the case of 6π-electrocyclization.

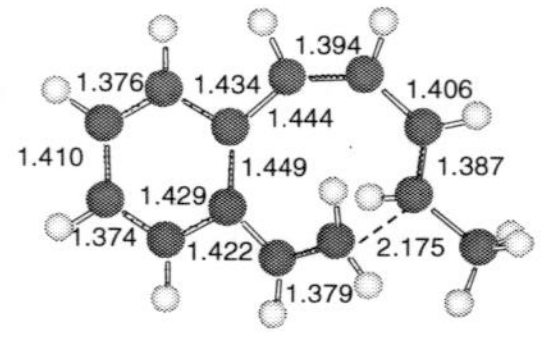

18a-TS

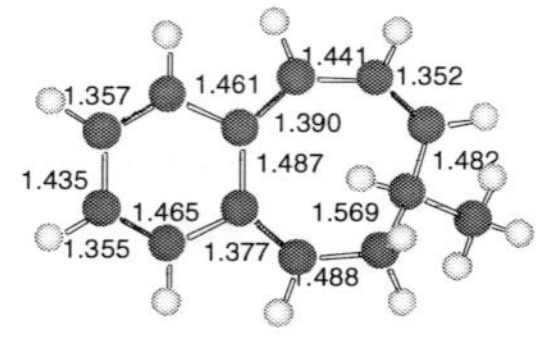

21a-TS

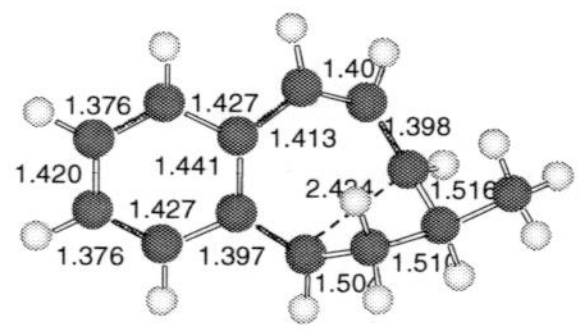

20a-TS

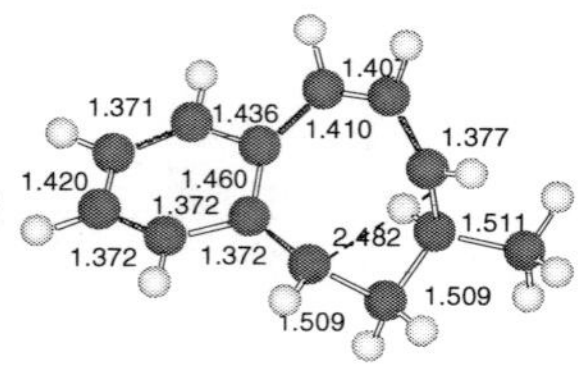

23a-TS

Figure 5. Transition state structures for 8π and 6π- electrocyclizations of **14a** optimized by M06/6-31+G(d,p) method (bond distances are given in Å).

The replacement of methyl group with phenyl and furyl substituents causes noticeable geometrical changes in transition state structures. The most interesting are bond lengths of the new forming C7···C8 bonds in transition states **18b-TS** and **18c-TS** (2.064 Å and 2.062 Å, respectively). These are significantly shorter than in the methyl case, and are associated with the change of electronic nature and conjugation of substituents attached to the terminal C7 carbon atom. In variance, bond lengths differences of the new forming C1···C6 bonds for phenyl and furyl substituted 6π-electrocyclization transition states relative to **20a-TS** and **23a-TS** are less pronounced (2.329, 2.346, 2.354, and 2.356 Å, for **20b-TS** (**20c-TS**) and **23b-TS** (**23b-TS**), respectively). Transition state structural similarity for 6π-electrocyclization could be the consequence of larger distance of substituents from the interacting carbon atom.

Computational results for thermal 8π,6π- electrocyclizations are in good accordance with experimental results which indicate that incorporation of octatetraene π-system in aromatic ring requires significantly higher temperatures for 8π- electrocyclization to take place than for parent octatetraene systems.

Relative activation energies for thermal 8π,6π-electrocyclization of **14a-c** (Table 1) showed that the M06/6-311+G(d,p)//M06/6-31+G(d,p) activation barriers for conformational interconversions are lower than the barriers for product formation.

Table 1. Activation energies (kcal/mol) for TSs 18, 20, 21 and 23[a]

Species	**a** (R=Me) (R=me)(R=methyl)	**b** (R=Ph) (R=phenyl)	**c** (R=furyl)
18(a-c)-TS	80.2	84.7	94.0
20(a-c)-TS	47.7	66.7	75.3
21(a-c)-TS	33.6	52.5	59.6
23(a-c)-TS	77.2	97.3	103.6

[a] M06/6-311+G(d,p)//M06/6-31+G(d,p)+ZPVE.

Under such circumstances, as dictated by the Curtin-Hammett principle, the ratio of the products is decided by the energies of the rate-limiting transition states leading to the products.

Energy barrier estimated for the formation of intermediate **19a** in thermal 8π-electrocyclization of **14a** *via* transition state **18a-TS** is 80.2 kJ/mol (Figure 6). This value is notably higher than calculated for the parent unsubstituted octatetraene at the same level of theory (54.4 kJ/mol). For an comparison, we estimated E_a at the M06-2X/6-311+G(d,p)//M06-2X/6-31+G(d,p)+ZPVE level is 54 kJ/mol), Houk [39] obtained 33.5 kJ/mol (at the MP2/6-31G(d) level), Cossio [40] estimated E_a is 70.3 kJ/mol (B3LYP/6-31G*), Schleyer [48] obtained 62.7 kJ/mol using RMP4SDTQ/6-31G*//RMP2(fu)//6-31G*+ZPVE (RHF/6-31G*) method and the experimentally determined E_a is 71.1 kcal/mol. [49] Our results are in accord with the effect of monosubstitution on the thermal 6*π*-electrocyclization of hexatrienes obtained by Houk showing that most of the substituents at C1 slightly increase the activation barrier of hexatriene electrocyclization, due to the steric crowding. [50] DFT studies by Fu (ONIOM(QCISD(T)/6-31+G(d,p)):B3LYP/6-311+G(2df,2p) method) are also in variance with our predictions, which showed that the monosubstitution at C1 does not exert any significant effect on the rates of 6π electrocylizations. [51]

Subsequent conformational change **19a→22a** *via* transition state **21a-TS** requires just 26.8 kJ/mol. This value implies smooth conformational interconversion. Intermediate cyclooctatriene **19a** could react in two ways. In a first, energetically favorable way, by 45.5 kJ/mol **19a** proceeds by 6π-electrocyclic process to the *endo*-**15a** product *via* transition state **20a-TS**. Electrocyclic ring opening process leading back to the reactant is energetically less likely by possessing the larger barrier of 59.3 kJ/mol. By feasible conformational change **19a→22a**, intermediate **19a** alters its conformation and forms less stable isomer **22a**. In the final step, the *exo*-**15a** product is obtained *via* transition state **23a-TS**. Energy profile indicates that activation energies enable formation of both *endo* and *exo* products. The reaction path which results in the *exo* product includes energetically costly conformational change and higher activation energy for the final 6π-electrocyclization, it is less favorable. These results are in good accord with the experimentally observed preference of the *endo*-**15a** product.

The reaction outcome for phenyl and furyl substituted octatetraenes **14b** and **14c** is different and only *endo* products *endo*-**15b(c)** were formed. There is a significant difference from the reaction energy profile of tetraene **14a** which could explain experimentally observed selectivities. Replacement of methyl by phenyl and furyl groups raises the activation barrier for the initial 8π-electrocyclization step. Similar activation energy increase was predicted by Cossio for C1-monosubstituted octatetranes. [40]

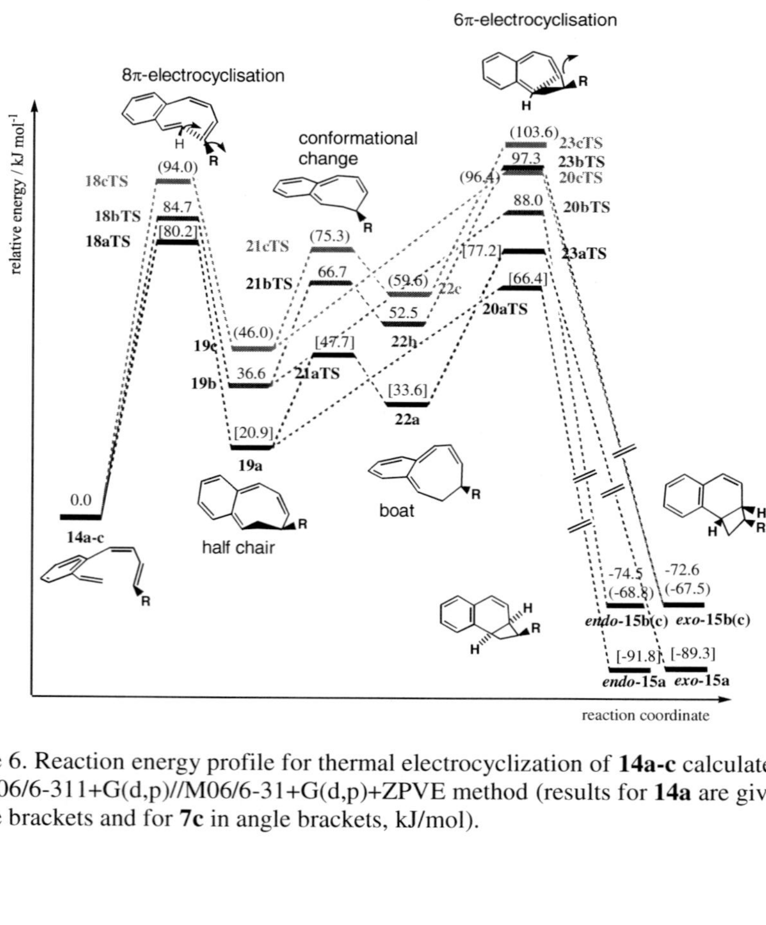

Figure 6. Reaction energy profile for thermal electrocyclization of **14a-c** calculated by the M06/6-311+G(d,p)//M06/6-31+G(d,p)+ZPVE method (results for **14a** are given in square brackets and for **7c** in angle brackets, kJ/mol).

Table 2. Gibbs free energy of solvation (ΔG_{solv}, kJ/mol)[a]

Species	**a** (R=Me) (R=methyl)	**b** (R=Ph) (R=phenyl)	**c** (R=furyl)
18(a-c)-TS	-5.4	-2.7	-4.9
20(a-c)-TS	-6.9	-4.5	-5.0
21(a-c)-TS	-6.3	-4.3	-4.9
23(a-c)-TS	-6.0	-4.1	-4.4

[a] CPCM/HF/6-31G(d)//M06/6-31+G(d,p) method.

In addition, activation barriers for 6π-electrocyclization leading to *endo/exo*-**15b(c)** products (*via* transition states **20b(c)-TS** and **23(c)-TS**) are similar or even higher than for 8π-electrocyclization **19b(c)-TS**. These differences indicate that energetically more accessible reaction **19b(c)→***endo-***15b(c)** produces the *endo* product and only the *endo* product is formed for phenyl and furyl derivatives. Activation energy higher than for 8π-electrocyclization (*via* **18b(c)-TS**) ultimately favors energetically lower electrocyclic ring opening process **19b(c)→14b(c)** and would revert to the reactants. **[52]**

The effects of solvent on the course of the studied reactions were estimated by employment of the CPCM/HF/6-31G(d) method in toluene (Table 2). Obtained Gibbs free energy of solvation (ΔG_{solv}) indicate that the *endo*-selectivity is retained, which is expected for an intramolecular reaction in nonpolar solvent.

Reactants are slightly destabilized upon solvation, whereas intermediates and products become stabilized upon solvation, but these small energy differences could not induce the reversal of the *endo*-selectivity.

3. Influence of Substituents in 2-Position on 8π,6π-Electrocyclizations of 2-Substituted [3,4]-Benzo-(*E,Z,Z,E*)-1,3,5,7-Octatetraenes [53]

Thermal electrocyclization reactions of 2-substituted benzooctatetraenes **24-26** give the *endo*-[4.2.0]bicyclooctadiene products **27** by 8π,6π-electrocyclization process (Scheme 8). In the case of chlorine substitution, this product further rearranges to give product **28**.

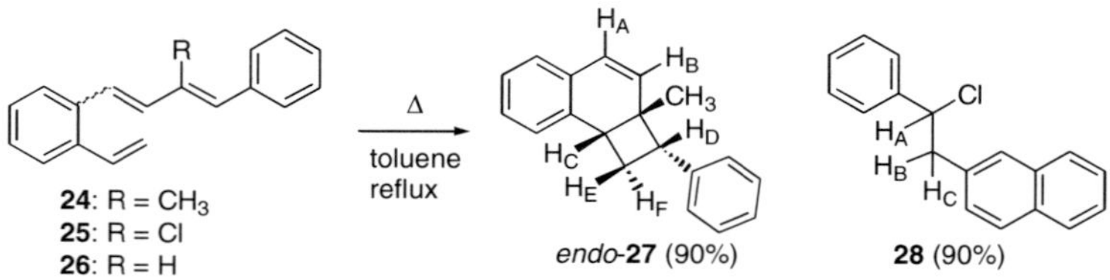

Scheme 8. Thermal reactions of benzooctatetraenes **24-26**.

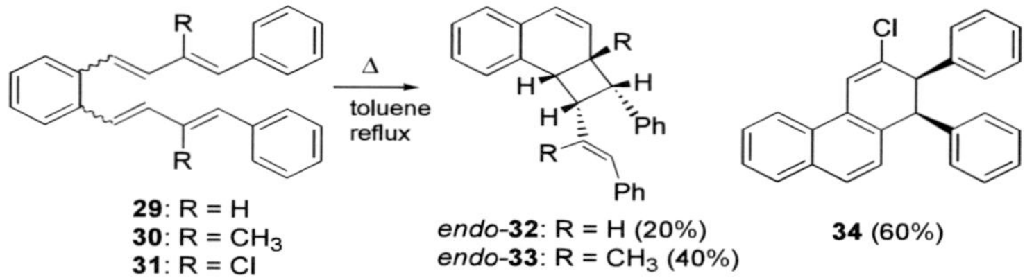

Scheme 9. Thermal reaction of benzodecapentaenes **29-31**.

Further extension of conjugated system to benzodecapentaenes **29-31** showed similar thermal reactivity as observed for **24** and **25** (Scheme 9) and thermolysis of mixtures of *cis,cis*-**29-31** and *cis,trans*-**29-31**, produced 8π,6π-electrocyclization products *endo*-**32** and *endo*-**33**, while initially formed chlorine product rearranged in these conditions to **34**.

The influence of substitution patterns on the electrocyclization processes of conjugated benzotetraenes and benzopentaenes, in which substituents are positioned at the C-2 of conjugated polyenes were studied at the M06/6-311+G(d,p)//M06/6-31G(d)+ZPVE level. In particular, the origins of *endo*-stereochemical outcomes of the thermal reactions of **24**, **25**, **29** and **30** and the activation barrier changes were computed.

Reaction mechanism for transformation of **24-26** to *endo*/*exo*-**27** is analogous to the one depicted in Scheme 7 for benzotetraenes **14a-c**. The introduction of methyl and chlorine substituents lowers the activation energy by 19 and 18 kJ/mol, respectively, which could be ascribed to activation of the tetraene system by electron donation.

Energy profile obtained for **24-26** (Figure 7) is very similar to the one found for **14a-c**. There is a relatively small energy barrier for the conformational change compared to electrocylization processes and the Curtin-Hammet principle dictates the *endo*/*exo* product ratio by the relative heights of the energy barriers. Subsequently the *endo*-**27** product is obtained *via* transition state **37-TS**, since the energy barrier is lower than the barrier for the formation of *exo*-**27** product (*via* transition state **40-TS**) regardless of substituent (by 10.5, 4.6 and 15.0 kJ mol^{-1}, for **36**, **37** and **38**, respectively) (Table 3). These results are in accord with the experimentally exclusive formation of the *endo*-**27** product. Benzodecapentaene system **29** possessing an additional olefinic bond behaves in similar manner as **26**, and *endo*-**32** product is preferred. For decapentaene **29**, the 8π electrocyclization process was computationally found to be more facile than 10π electrocyclization.

Figures 8 and 9 depict transition state structures for 8π and 6π electrocyclizations of **24-26** and **29** optimized by M06/6-31+G(d,p) method. The geometrical features of transition state strucutres for 8π electrocyclizations feature C...C separation involving the C-C bond being formed. These values are within the range of 2.032-2.119 Å, and are typical C...C separations (ca. 2.15±0.15 Å) computed for other pericyclic reactions. [39]

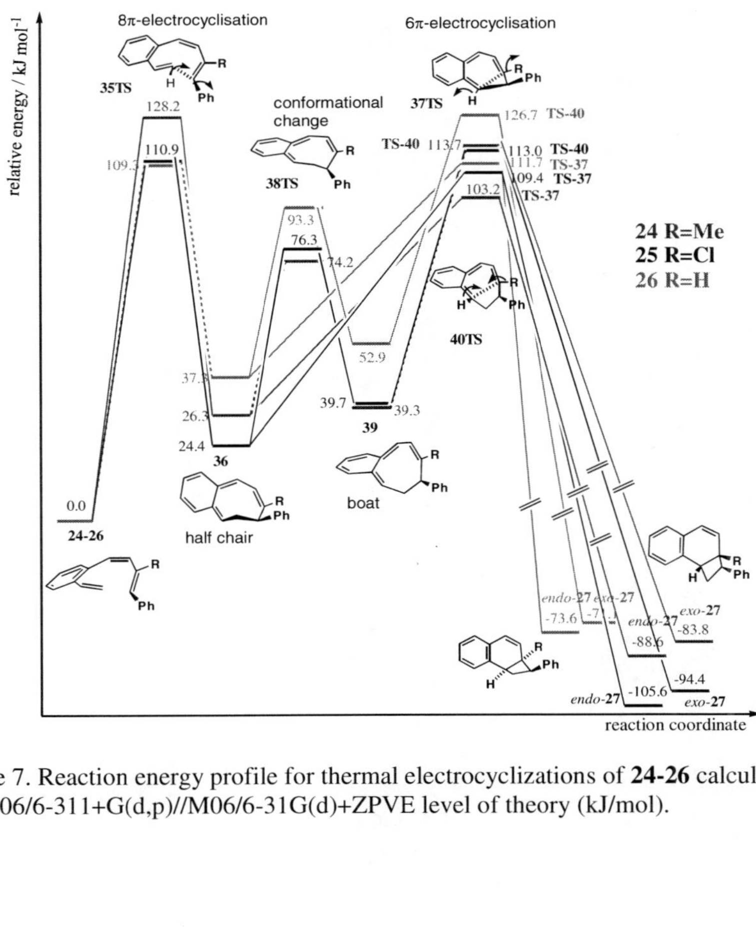

Figure 7. Reaction energy profile for thermal electrocyclizations of **24-26** calculated at the M06/6-311+G(d,p)//M06/6-31G(d)+ZPVE level of theory (kJ/mol).

Table 3. Activation energies (kcal/mol) for TSs 35, 38, 37 and 40[a]

Species	**1** (R=Me) (R=methyl)	**2** (R=Cl)	**3** (R=H)
35-TS	109.2	110.9	128.2
38-TS	80.5	76.3	93.3
37-TS	103.2	109.4	111.7
40-TS	113.7	113.0	126.7

[a] M06/6-311+G(d,p)//M06/6-31+G(d,p)+ZPVE.

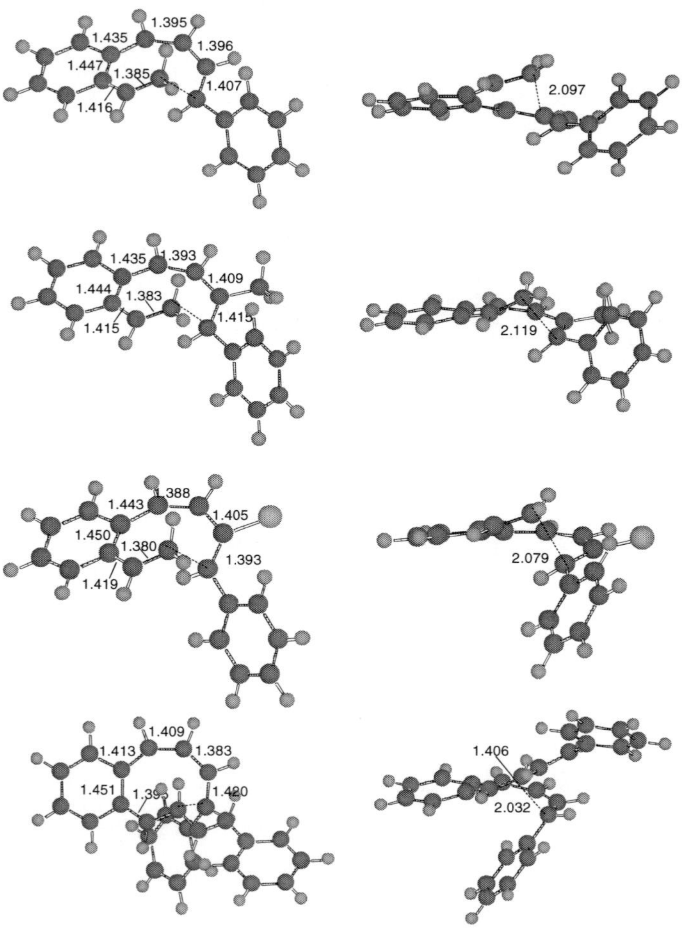

26 H

24 Me

25 Cl

29

Figure 8. Transition state structures for 8π electrocyclizations of **24-26** and **29** optimized by M06/6-31+G(d,p) method (bond distances are given in Å).

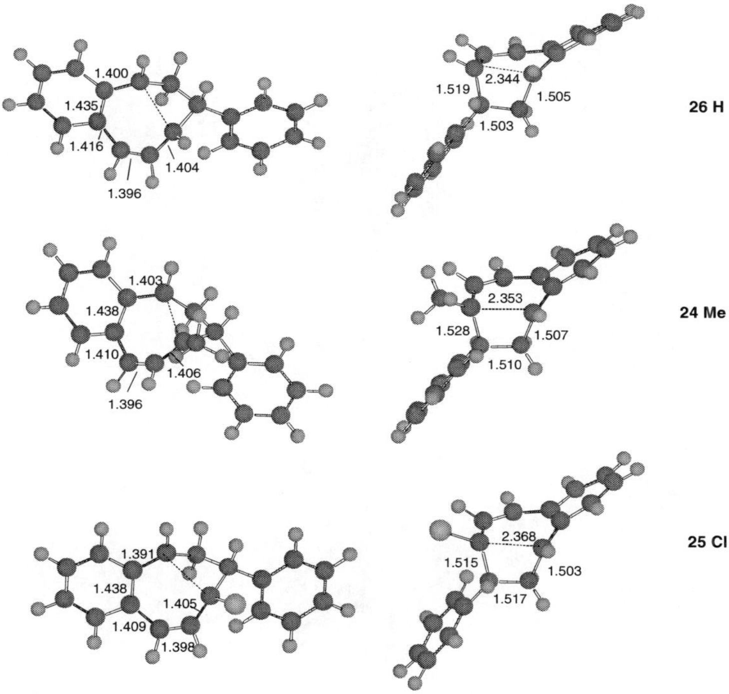

Figure 9. Transition state structures for 6π electrocyclizations of **24-26** optimized by M06/6-31+G(d,p) method (bond distances are given in Å).

4. Influence of Aromatic Ring on 8π,6π-Electrocyclizations of [3,4]-Aryl-(*E*,*Z*,*Z*,*E*)-1,3,5,7-Octatetraenes [54]

Computational predictions (employing the M06-2X/6-311+G(d,p)//M06-2X/6-31G(d)+ZPVE method) on the influence of aromaticity on the activation barriers for 8π,6π-electrocyclizations were carried out for a series of biphenyls listed in Table 4. Geometry and frontier orbital symmetry of the terminal double bonds of 2-hydroxy-dibenzooctatetraene for correct orbital overlap for 8π electrocyclization is shown in Figure 10a. Orbital analysis reveals that aromatic moieties of aryl rings are involved in 8π-tetraene system. Similar contribution of aryl rings π-orbitals was found for the concominant 6π-electrocyclization (Figure 11).

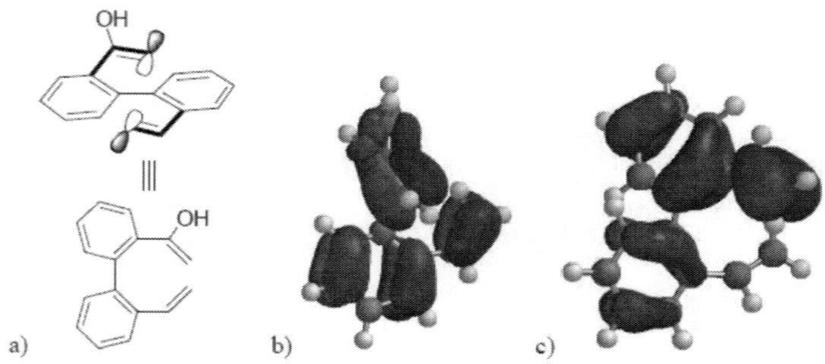

Figure 10. a) Required geometry of the terminal double bonds of 2-hydroxy-dibenzo octatetraene for correct orbital overlap for 8π electrocyclization, b) depiction of the HOMO-1 orbital and c) HOMO of dibenzooctatetraene (RHF/6-31G*).

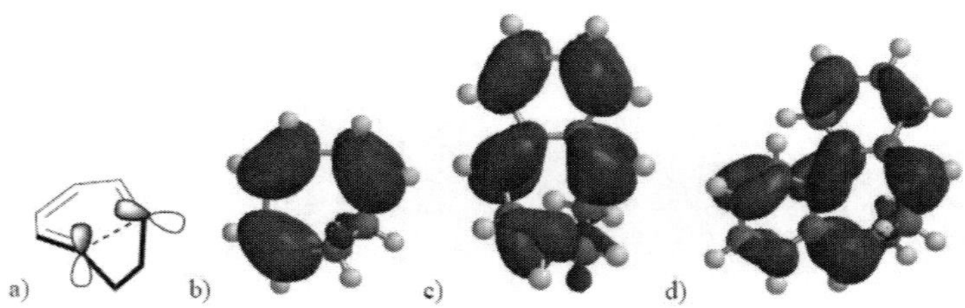

Figure 11. a) Required geometry of the terminal double bonds of octatriene intermediate for correct orbital overlap in 6π electrocyclization, b) depiction of the HOMO orbital octatriene (RHF/6-31G*), c) HOMO of benzooctatriene and d) HOMO of dibenzooctatriene.

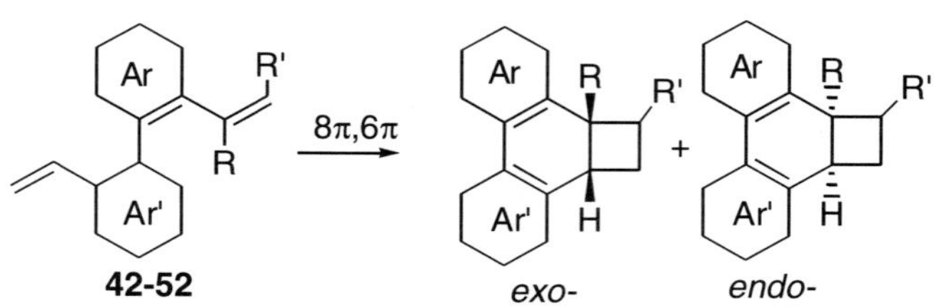

Activation energies obtained for a series of [3,4]-aryl-(*E,Z,Z,E*)-1,3,5,7-octatetraenes **42-52** suggest that the incorporation of the second aromatic ring in the octatetraene moiety of **42** significantly raises E_a for the 8π electrocyclization process in **43** (from 84.0 to 160.8 kJ/mol). These computational results are in good accordance with a prediction of Takasu *et al.* [55] that 8π electrocyclization of biaryls **42** bearing acyl and alkenyl substituents at the 2 and 2' positions, leading to a doubly dearomatized intermediate, would require a high activation energy.

Table 4. Activation energies for 8π and 6π electrocyclizations of [3,4]-aryl-(*E,Z,Z,E*)-1,3,5,7-octatetraenes 42-52 (kJ/mol) obtained at the level [a,b]

	Ar'	Ar	R	R'	E_a TS (8π)	E_a TS (conf)	E_a TS (6π)
41					43.4		108.6
42	Ph				84.0		53.5 *endo*
						27.2	48.0 *exo*
43	Ph	Ph			160.8		6.5 exo
44	Ph		OH		81.6		49.8 *exo*
						26.2	40.0 *endo*
45	Ph	Ph	OH		158.9	0.6	6.1 *endo*
46	Ph		OH	Ph	86.8		41.8 *exo*
						35.1	33.3 *endo*
47	Naphthyl				102.4		35.7 *endo*
						53.7	29.8 *exo*
48	Phenanthryl				101.0		88.7 *endo*
						80.9	90.6 exo
49	2'-pyridyl				71.7		55.5 *endo*
						28.7	49.8 *exo*
50	2'-furyl				37.8		78.6 *exo*
51	2'-thioph				50.8		72.8 *exo*
52	2'pyrrolyl				58.7		87.0 *exo*

[a] Relative to the thermodynamically most stable octatetraene isomer; [b] M062x/6-311+G**// M062x/6-31G* + ZPVE energies.

Addition of hydroxy group in the position 2 of octatetraene moiety has just a minor influence on the computed E_a, which is in accord with the small influence of substituent at C2 obtained by Fu. [51] He fund that C2 substituent will decrease E_a of hexatriene 6π electrocyclization, probably by the anchimeric assistance effect exerted by the substituent on the transition state. [56] An electron acceptor tends to bring about more significant decrease in the activation free energy.

Furthermore, higher activation energies are obtained for aromatic substituents which possess more aromatic character. On the other hand, more aromatic rings substantially decrease the activation energies required for 6π electrocyclization process, as the favorable proces to regain aromatic stability. Interesting is the inclusion five-membered heteroaromatic moieties, for which DFT calculations indicate lower activation energies in comparison to benzene ring. Products of the 8π electrocyclizations are very stable (even more stable than the starting octatetraenes), and therefore activation energies for the 6π electrocyclization are higher than for 8π process.

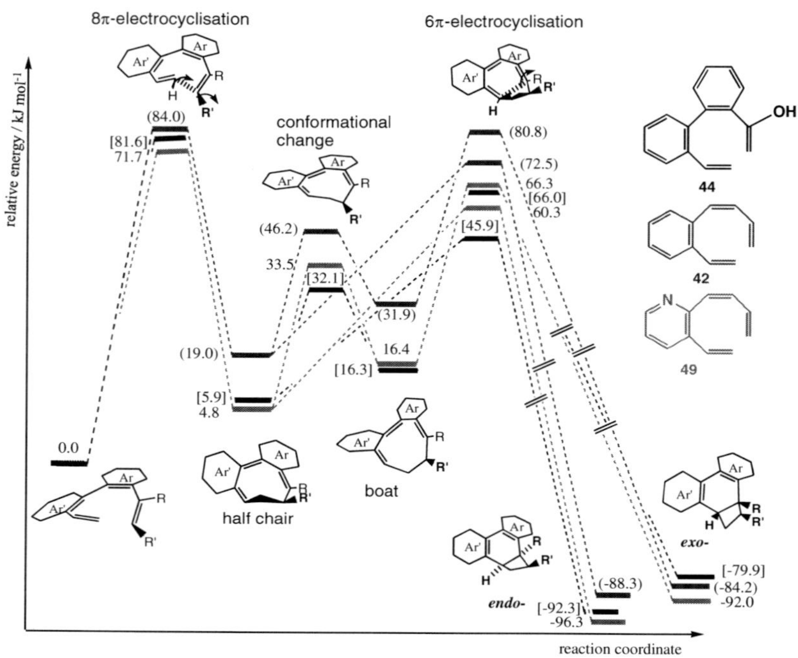

Figure 12. Reaction energy profile for thermal electrocyclizations of **42**, **44** and **49** calculated at the M06-2X/6-311+G(d,p)//M06-2X/6-31G(d)+ZPVE level of theory (kJ/mol).

For the furan-embedded octatetraene **50**, the lowest E_a was calculated (37.8 kJ/mol), which is value lower than for the parent cyclooctatetraene, suggesting almost spontaneous 8π electrocyclization. Energy profiles for 8π,6π- electrocyclization cascade for [3,4]-aryl-(*E,Z,Z,E*)-1,3,5,7-octatetraenes are illustrated in Figure 12 for octatetraenes **42**, **44** and **49**. These profiles are very similar to those described earlier in Figures 6 and 7.

Geometrical features of transition state strucutres for 8π and 6π-electrocyclizations of aromatic tetraenes are illustrated in Figure 13 and summarized in Table 5. The C...C separation involving the C-C bond being formed in 8π electrocyclic process are within the range of 1.858-2.206 Å, and are typical C...C separations (ca. 2.15±0.15 Å) computed for other pericyclic reactions.[39] The bond lengths around the periphery of octatetratraene, expressed as the maximum difference between the longest and shortest bond (Δr, excluding the forming bond itself), reveal the small degree of bond alternation typical of aromatic ring transition states (Δr=0.041-0.103 Å) [57], indicating aromatic transition states. [58]

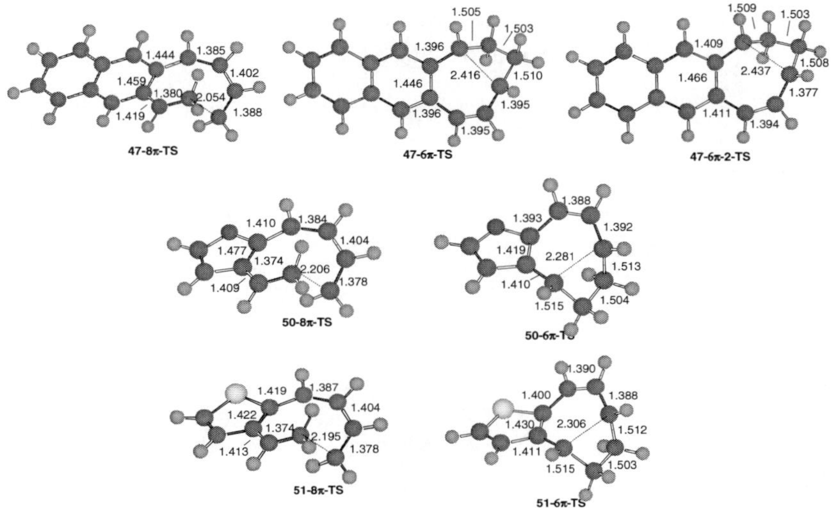

Figure 13. Transition state structures for 8π and 6π- electrocyclizations of tetraenes **47**, **50** and **51** optimized by M06-2X/6-31G* method (bond distances are given in Å).

The transition structure can be termed aromatic since the C-C bond lengths are nearly equal and comparable to that in benzene. Besides the geometrical criterion of aromaticity, a magnetic criterion can be employed, the NICS(0) magnetic index [59] Caclulated NICS(0) values (B3LYP/6-311+G** calculations) for 8π transition structures for **44**, **45**, and **43** indicate substantial Möbius aromaticity of TSs: -9.79 ppm (*zz*=-2.63), -6.09 ppm (*zz*=-5.94) and -10.31 ppm (*zz*=-2.01), respectively. These NICS(0) values are similar to the one obtained by Schleyer [60] for 8π electrocyclization of *Z,E,Z*-decapentaene (-9.8 ppm), or by Yamaguchi [62], calculated at the B3LYP/6-311+G(d,p) level (-9.8 ppm). NICS(0) magnetic index for 10π electrocyclization has a typically highly aromatic value (-14.9 ppm, cf. -10 ppm for the benzene), [57] while the 6*π* electrocyclization of hexatetraene has NICS value -10.38 ppm. [61]

Analysis of geometries features of TS structures showed that the aromatic C=C bond lenghts involved in TS can not be directly correlated to their aromaticity.

The C...C separation involving the C-C bond being formed in the second, 6π electrocyclic process are by average longer and fall within the range of 2.226-2.586 Å.

Table 5. TS geometries for 8π and 6π- electrocyclizations, aromatic NICS values

	d(C···C) (Å)	Δr (Å)	d(Arene C-C) (Å)	d(C···C) (Å)	Δr (Å)	d(Arene C-C) (Å)
41	2.329	0.050	-	2.172	0.021	-
42	2.128	0.068	1.444	2.351	0.069	1.451
43	1.858	0.041	1.446	2.568	0.084	1.443
44	2.166	0.070	1443	2.340	0.064	1.445
45	2.093	0.048	1.442	2.586	0.066	1.440
46	2.086	0.079	1.459	2.380	0.046	1.434
47	2.054	0.079	1.459	2.416	0.051	1.446
48	2.134	0.070	1.417	2.226	0.048	1.434
49	2.140	0.066	1.441	2.325	0.041	1.435
50	2.206	0.103	1.477	2.281	0.031	1.419
51	2.195	0.048	1.422	2.306	0.040	1.430
52	2.175	0.056	1.432	2.306	0.046	1.434

Computational predictions on the influence of aromaticity on the activation barriers for π electrocyclizations are in accord with the recent literature example of Yamaguchi. [62] He reported that the 8π electrocyclic reaction was facilitated by a less aromatic heteroarene(thiazole)-tetraene system. 6π electrocyclization does not follow, since rearomatization is much more facile reaction which leads to the formation of the energetically stable cyclooctatetraene products. A B3LYP/6-311G(d,p) quantum chemical study showed that for the 8π electrocyclic reaction change of terminal substituent from Br (instead of H) does not influence activation energies at all (29.0 vs. 31.2 kcal/mol).

Our results on the influence of aromatic moiety on the activation barriers for electrocyclizations could be also correlated with the report by Zora. [63] His (U)B3LYP/6-31G* calculations on the influence of polyene system part of aromatic ring at the termini on the 6π electrocyclizations of dienylketenes to 2,4-cyclohexadienones showed that if the terminal double bond is embedded into a benzenoid-type aryl moiety, the partial or complete loss of aromaticity increases the activation barrier. The effect of aromaticity is slightly less pronounced for dienylketenes carrying five-membered heterocyclic aromatic substituents.

Finally, our computational findings are in good accord with aromaticity influence on the peri selectivity of the electrocyclization of 1-aryl- and heteroaryl-substituted (1*Z*,3*Z*)-1,3,5-hexatrienes. DFT calculations (B3LYP/6-31G*) carried out by Saá [64] indicated that for Ar=Ph 8π E_a=22.0 kcal/mol, and 6π E_a=28.0 kcal/mol, while for indolo activation energy lowers for 8π E_a=14.4 kcal/mol, whereas for 6π E_a=28.9 kcal/mol. In Saá's report 8π electrocyclizations are kinetically, whereas 6π electrocyclizations are thermodynamically favored. In the case of benzene derivative, the loss of aromaticity of the benzene ring involved in the 8π electrocyclization makes this process highly endothermic, with the competitive 6π electrocyclization more favored. Conversely, if the 8π electrocyclization involves a heteroaryl ring, *e.g.*, indole, the process becomes exothermic with lower activation barrier, therefore making this process more favorable.

CONCLUSION

Besides of their theoretical and mechanistic insterest, the 8π,6π electrocyclization cascade presents an biosyntheticaly important process. Quantum-chemical studies at the state of the art computational level help us to reveal the detailed mechanistic insight into this cascade reaction.

REFERENCES AND NOTES

[1] Rappoport, Z., Ed., *The Chemistry of Dienes and Polyenes*, Volume 2, Wiley, Chichester, 2000.

[2] Burnley, J.; Ralph, M.; Sharma, P.; Moses, J. E. in *Biomimetic Organic Synthesis*, Poupon, E.; Nay, B., Eds., Vol. 1. Wiley, Weinheim, 2011, pp 591-636.

[3] Poupon, E.; Nay, B. (Eds.), *Biomimetic Organic Synthesis*. Wiley, Weinheim, 2011.

[4] Oxaelectrocyclization: Fotiadou, A. D.; Zografos, A. L. Electrocyclization of oxatrienes in the construction of structurally complex pyranopyridones. *Org. Lett.* 2012, *16*, 5664-5667.

[5] Azaelectrocyclization: Cheng, X.; Waters, S. P. Pyridone Annulation via Tandem Curtius Rearrangement/6π-Electrocyclization: Total Synthesis of (-)-Lyconadin C. *Org. Lett.* 2013, *15*, 4226-4229.

[6] Carruthers, W.; Coldham, I., *Modern Methods of Organic Synthesis*, Cambridge University Press, Cambridge 2004.

[7] Bandaranayake, W. M.; Banfield, J. E.; Black, D. St. C. Postulated electrocyclic reactions leading to endiandric acid and related natural products. *J. Chem. Soc. Chem. Commun.* 1980, 902-903.

[8] Kim, K.; Lauher, J. W.; Parker, K. A. Asymmetric Induction in 8π Electrocyclizations. Design of a Removable Chiral Auxiliary. *Org. Lett.* 2012, *14*, 138-141.

[9] Thompson, S.; Coyne, A. G., Knipe, P. C.; Smith, M. D. Asymmetric electrocyclic reactions. *Chem. Soc. Rev.* 2011, 40, 4217-4231.

[10] a) Miller, A. K.; Trauner, D. Mapping the Chemistry of Highly Unsaturated Pyrone Polyketides. *Synlett* 2006, *14*, 2295-2316; b) Beaudry, C. M.; Malerich, J. P.; Trauner, D. Biosynthetic and Biomimetic Electrocyclizations. *Chem. Rev.* 2005, *105*, 4757-4778.

[11] Huisgen, R.; Dahmen, A.; Huber, H. Stereospecific conrotatory valence isomerization of octatetraenes to cycloocta-1,3,5-trienes. *J. Am. Chem. Soc.* 1967, *89*, 7130-7131.

[12] Nicolau, K. C. Inspirations, Discoveries, and Future Perspectives in Total Synthesis. J. Org. Chem. 2009, 74, 951-972.

[13] Bandaranayake, W. M.; Banfiled, J. E.; Black, D. St. C.; Fallon, G. D.; Gatehouse, B. M. Endiandric acid, a novel carboxylic acid from *Endiandra introrsa* (Lauraceae): *X*-ray structure determination *J. Chem. Soc. Chem. Commun.* 1980, 162-163.

[14] Parker, K. A.; Lim, Y.-H., The Total Synthesis of (-)-SNF4435 C and (+)-SNF4435 D. *J. Am. Chem. Soc.* 2004, *126*, 15968-15969.

[15] Moses, J. E.; Baldwin, J. E.; Marquez, R.; Adlington, R. M.; Cowley, A. R. Studies on the Biomimetic Synthesis of SNF4435 C and D. *Org. Lett.* 2002, *4*, 3731-3734.

[16] a) Jacobsen, M. F.; Moses, J. E.; Adlington, R. M.; Baldwin, J. E. The Total Synthesis of Spectinabilin and Its Biomimetic Conversion to SNF4435C and SNF4435D. *Org. Lett.* 2005, *7*, 2473-2476; b) Jacobsen, M. F.; Moses, J. E.; Adlington, R. M.; Baldwin, J. E. The biomimetic synthesis of SNF4435C and SNF4435D, and the total synthesis of the polyene metabolites aureothin, *N*-acetyl-aureothamine and spectinabilin. *Tetrahedron* 2006, *62*, 1675-1689.

[17] Beaudry, C. M.; Trauner, D. Synthetic Studies toward SNF4435 C and SNF4435 D. *Org. Lett.* 2002, *4*, 2221-2224.

[18] Nicolaou, K. C.; Petasis, N. A.; Uenishi, J.; Zipkin, R. E. The Endiandric Acid Cascade. Electrocyclizations in Organic Synthesis. 1. Stepwise, Stereocontrolled Total Synthesis of Endiandric Acids A and B. *J. Am. Chem. Soc.* 1982, *104*, 5555-5557.

[19] Nicolaou, K. C.; Petasis, N. A.; Zipkin, R. E. The endiandric acid cascade. Electrocyclizations in organic synthesis. 4. Biomimetic approach to endiandric acids A-G. Total synthesis and thermal studies. *J. Am. Chem. Soc.* 1982, *104*, 5560-5562.

[20] a) Rickards, R. W.; Skropeta, D. Electrocyclic processes in aromatic biomimetic study of pseudorubrenoic acid A. *Tetrahedron* 2002, *58*, 3793-3800; b) Skropeta, D.; Rickards, R. W. Domino pericyclic reactions of acyclic conjugated (*E,Z,E,E*)-tetraenes. *Tetrahedron* 2007, *48*, 3281-3284.

[21] Parker, K. A.; Lim, Y.-H. "*Endo*" and "*Exo*" Bicyclo[4.2.0]-octadiene Isomers from the Electrocyclization of Fully Substituted Tetraene Models for SNF 4435C and D. Control of Stereochemistry by Choice of a Functionalized Substituent. *Org. Lett.* 2004, *6*, 161-164.

[22] Parker, K. A.; Wang, Z. Cleavable Chiral Auxiliaries in 8π (8π, 6π) Electrocyclizations. *Org. Lett.* 2006, *8*, 3553-3556.

[23] Marvell, E. N.; Seubert, J.; Vogt, G.; Zimmer, G.; Moy, G.; Siegmann, J. R. Electrocyclic reactions of tetraenes: Influence of terminal substituents. *Tetrahedron* 1978, *34*, 1307-1323.

[24] Salem, B.; Suffert, J. A 4-*exo*-dig Cyclocarbopalladation/8π Electrocyclization Cascade: Expeditious Access to the Tricyclic Core Structures of the Ophiobolins and Aleurodiscal. *Angew. Chem. Int. Ed.* 2004, *43*, 2826-2830.

[25] Hayashi, R.; Fernández, S.; Okamura, W. H. An 8π Electron Electrocyclization Leading to a 9,19-Methano-Bridged Analogue of 1α,25-Dihydroxyvitamin D_3. *Org. Lett.* 2002, *4*, 851-854.

[26] Paquette, L. A.; Tae, J. 1,5-Asymmetric Induction in Squarate Cascades. Conformational Control of Helicity by Chiral Amino Substituents during Conrotatory Octatetraene Cyclization Prior to β-Elimination. *J. Org. Chem.* 1998, *63*, 2022-2030.

[27] Miller, A. K.; Trauner, D. Mining the Tetraene Manifold: Total Synthesis of Complex Pyrones from *Placobranchus ocellatus*. *Angew. Chem. Int. Ed.* 2005, *44*, 4602-4606.

[28] Barbarow, J. E.; Miller, A. K.; Trauner, D. Biomimetic Synthesis of Elysiapyrones A and B. *Org. Lett*. 2005, *7*, 2901-2903.

[29] Moses, J. E.; Adlington, R. M.; Rodriguez, R.; Eade, S. J.; Baldwin, J. E. Biomimetic synthesis of (±)-9,10-deoxytridachione. *Chem. Commun.* 2005, 1687-1689.

[30] Manzo, E.; Ciavatta, M. L.; Gavagnin, M.; Mollo, E.; Wahidulla, S.; Cimino, G. New γ-pyrone propionates from the Indian Ocean

sacoglossan *Placobranchus ocellatus*. *Tetrahedron Lett.* 2005, *46*, 465-468.

[31] Sofiyev, V.; Navarro, G.; Trauner, D. Biomimetic Synthesis of the Shimalactones. *Org. Lett.* 2008, *19*, 149-152.

[32] Sharma, P.; Ritson, D. J.; Burnley J.; Moses, J. E. A synthetic approach to kingianin A based on biosynthetic speculation. *Chem. Commun.* 2011, *47*, 10605-10607.

[33] Beaudry, C. M.; Trauner. D. Total Synthesis of (-)-SNF4435 C and (+)-SNF4435 D. *Org. Lett.* 2005, *7*, 4475-4477.

[34] Marvel, E. N. *Thermal Electrocyclic Reactions*, Academic Press, New York, 1980.

[35] Woodward, R. B.; Hoffmann, R. *The Conservation of Orbital Symmetry*, Verlag Chemie, Weinheim, 1970.

[36] Oliva, M.; Domingo, L. R.; Safont, V. S.; Andrés, J.; Castillo, R.; Moliner, V. Theoretical study of the molecular mechanism of the domino pathways for squarate ester sequential reactions. *J. Phys. Org. Chem.* 1999, *12*, 61-69.

[37] Pichko, V. A.; Simkin, B. Y.; Minkin, V. I. Conrotatory and disrotatory reaction paths for thermal and photoinduced ring-closing reactions of 1,3,5-hexatriene and its isoelectronic analogs. *J. Org. Chem.* 1992, *57*, 7087-7092.

[38] Thomas, B. E.; Evanseck, J. D.; Houk, K. N. Electrocyclic reactions of 1-substituted 1,3,5,7-octatetraenes. An ab initio molecular orbital study of torquoselectivity in eight-electron electrocyclizations *J. Am. Chem. Soc.* 1993, *115*, 4165-4169.

[39] Houk, K. N.; Li, Y.; Evanseck, J. D. Transition Structures of Hydrocarbon Pericyclic Reactions. *Angew. Chem. Int. Ed.* 1992, *31*, 682-708.

[40] Lecea, B.; Arrieta, A.; Cossio, F. P. Substituent Effects in Eight-Electron Electrocyclic Reactions. *J. Org. Chem.* 2005, *70*, 1035-1041.

[41] Hulot, C.; Amiri, S.; Blond, G.; Schreiner, P. R.; Suffert, J. Understanding the Torquoselectivity in 8π-Electrocyclic Cascade Reactions: Synthesis of Fenestradienes versus Cyclooctatrienes. *J. Am. Chem. Soc.* 2009, *131*, 13387-13398.

[42] Škorić, I.; Pavošević, F.; Marinić, Ž.; Šindler, M.; Eckert-Maksić, M.; Vazdar, M.; Margetić, D. Thermal reaction of [3,4]-benzo-8-methyl-(*E,Z,Z,E*)-1,3,5,7-octatetraenes and quantum-chemical study of the (8π,6π)-electrocyclisation. *Org. Biomol. Chem.* 2011, *9*, 6771-6778.

[43] Chang, M.-Y.; Wu, M.-H.; Chen, Y.-L. One-Pot Synthesis of Substituted Tetrahydrocyclobuta[*a*]naphthalenes by Domino Aldol Condensation/Olefin Migration/Electrocyclization. Org. Lett. 2013, *15*, 2822-2825.

[44] Dahmen, A.; Huisgen, R. Zur stereospezifität der conrotatorischen cyclisierung des decatetraens. *Tetrahedron Lett.* 1969, *10*, 1465-1469.

[45] Zhao, Y.; Truhlar, D. G. The M06 suite of density functionals for main group thermochemistry, thermochemical kinetics, noncovalent interactions, excited states, and transition elements: two new functionals and systematic testing of four M06-class functionals and 12 other functionals. *Theor. Chem. Acc.* 2008, *120*, 215-241.

[46] Zhao, Y.; Truhlar, D. G. Density functionals with broad applicability in chemistry. *Acc. Chem. Res.* 2008, *41*, 157-167.

[47] Um, J. M.; Xu, H.; Houk, K. N.; Tang, W. Thermodynamic Control of the Electrocyclic Ring Opening of Cyclobutenes: C=X Substituents at C-3 Mask the Kinetic Torquoselectivity. *J. Am. Chem. Soc.* 2009, *131*, 6664-6665.

[48] Jiao, H.; Schleyer, P. v. R. Evidence for the Möbius Aromatic Character of Eight π Electron Conrotatory Transition Structure. Magnetic Criteria. *J. Chem. Soc. Perkin Trans. 2* 1994, 407-410.

[49] Goldfarb, T. D.; Lindquist, L. J. Flash photolysis studies of 1,3,5-cyclooctatriene Reversible ring opening. *J. Am. Chem. Soc.* 1967, *89*, 4588-4592.

[50] Evanseck, J. D.; Thomas, B. E., IV; Spellmeyer, D. C.; Houk, K. N. Transition Structures of Thermally Allowed Disrotatory Electrocyclizations. The Prediction of Stereoselective Substituent Effects in Six-Electron Pericyclic Reactions. *J. Org. Chem.* 1995, *60*, 7134-7141.

[51] Yu, T.-Q.; Fu, Y.; Liu, Y.; Guo, Q.-X. How to Promote Sluggish Electrocyclization of 1,3,5-Hexatrienes by Captodative Substitution. J. Org. Chem., 2006, *71*, 6157-6164.

[52] Predictions obtained by the BMK/6-311+G(d,p)//BMK/6-31G(d)+ZPVE method are essentially identical, while the MP2/6-311+G(d,p)//MP2/6-31G(d)+ZPVE method, which tends to underestimate barrier heights for pericyclic reactions (Thomas, J. B.; Waas, J. R.; Harmata, M.; Singleton, D. A. Control Elements in Dynamically Determined Selectivity on a Bifurcating Surface. *J. Am. Chem. Soc.* 2008, *130*, 14544-14555.) gave incorrect predictions of stereochemical outcomes.

[53] Vuk, D.; Marinić, Ž.; Molčanov, K.; Margetić, D.; Škorić, I. Thermal electrocyclisation reactions II: Benzooctatetraenes and benzodecapentaenes. *Tetrahedron* 2014, *70*, 886-891.

[54] Jušinski, I.; Margetić, D. *Croatica Chem. Acta*, 2014, paper in preparation.

[55] Nagamoto, Y.; Yamaoka, Y.; Fujimura, S.; Takemoto, Y.; Takasu, K. *Org Lett.* 2014, *16*, 1008-1011.

[56] Spangler, C. W.; Ibrahim, S.; Bookbinder, D. C.; Ahmad, S. Thermal electrocyclic ring closure of alkenylhexa-1,3,5-trienes and the question of anchimeric π-bond participation. *J. Chem. Soc., Perkin Trans.* 1979, 717-719.

[57] Rzepa, H. S. Double-twist Möbius aromaticity in a 4n+2 electron electrocyclic reaction. *Chem. Commun.* 2005, 5220-5222.

[58] Jiao, H.; Schleyer, P. v. R. A Detailed Theoretical Analysis of the 1,7-Sigmatropic Hydrogen Shift: The Möbius Character of the Eight-Electron Transition Structure. *Angew. Chem., Int. Ed. Engl.* 1993, *32*, 1763-1765.

[59] Schleyer, P. v. R.; Maerker, C.; Dransfeld, A.; Jiao, H.; van Eikema Hommes, N. J. Nucleus-Independent Chemical Shifts: A Simple and Efficient Aromaticity Probe. *J. Am. Chem. Soc.* 1996, *118*, 6317-6318.

[60] Schleyer, P. v. R.; Judy I. Wu, J. I.; Cossío, F. P.; Fernández, I. Aromaticity in transition structures. *Chem. Soc. Rev.* 2014, Advance Article, DOI: 10.1039/C4CS00012A

[61] Arrieta, A.; Abel de Cozar, A.; P. Cossio, F. P. Cyclic Electron Delocalization in Pericyclic Reactions, *Curr. Org. Chem.* 2011, *15*, 3594-3608.

[62] Mouri, K.; Saito, S.; Hisaki, I.; Yamaguchi, S. Thermal 8π electrocyclic reaction of heteroarene tetramers: new efficient access to π-extended cyclooctatetraenes. *Chem. Sci.* 2013, 4, 4465-4469.

[63] Zora, M. Transition Structures, Energetics, and Nucleus-Independent Chemical Shifts for 6π Electrocyclizations of Dienylketenes to Cyclohexadienones: A DFT Study. J. Org. Chem. 2004, *69*, 1940-1947.

[64] García-Rubín, S.; Varela, J. A.; Castedo, L.; Saá, C. 6πe- versus 8πe-Electrocyclization of 1-Aryl- and Heteroaryl-Substituted (1*Z*,3*Z*)-1,3,5-Hexatrienes: A Matter of Aromaticity, Org. Lett. 2009, *11*, 983-986.

INDEX

A

B

E

F

G

H

I

J

K

L

M

N

O

P

Q

R

S

T

U

V

W

Y

Z